Freeman Laboratory Separates in Biology

The exercises listed below are available as Freeman Laboratory Separates. Each Separate is derived from the experiments in the fifth edition of this manual. The Separates are self-bound, self-contained exercises. They are 8½ inches by 11 inches in size, and are punched for a three-ring notebook. They can be ordered in any assortment or quantity. Order through your bookstore or call our Order Department at 1-800-221-7945, specifying the ISBN and title.

LABORATORY SEPARATES BY WALKER AND HOMBERGER

From *Anatomy and Dissection of the Fetal Pig*, Fifth Edition
1998, 140 pages, ISBN 0-7167-2637-8

0-7167-9296-6 External Anatomy, Skin, and Skeleton
0-7167-9297-4 Muscles
0-7167-9298-2 Digestive and Respiratory Systems
0-7167-9299-0 Circulatory System
0-7167-9300-8 Urogenital System
0-7167-9301-6 Nervous Coordination: Sense Organs
0-7167-9302-4 Nervous Coordination: Nervous System
0-7167-9288-5 Glossary of Vertebrate Anatomical Terms

From *Anatomy and Dissection of the Rat*, Third Edition
1998, 140 pages, ISBN 0-7167-2635-1

0-7167-9289-3 External Anatomy, Skin, and Skeleton
0-7167-9290-7 Muscles
0-7167-9291-5 Digestive and Respiratory Systems
0-7167-9292-3 Circulatory System
0-7167-9293-1 Urogenital System
0-7167-9294-X Nervous Coordination: Nervous System
0-7167-9295-8 Nervous Coordination: Sense Organs
0-7167-9288-5 Glossary of Vertebrate Anatomical Terms

From *Dissection of the Frog*, Second Edition
1998, 96 pages, ISBN 0-7167-2636-X

0-7167-9218-4 External Anatomy, Skin, and Skeleton
0-7167-9221-4 Circulatory System
0-7167-9220-6 Digestive and Respiratory Systems
0-7167-9219-2 Muscles
0-7167-9223-0 Nervous Coordination: Nervous System
0-7167-9224-9 Nervous Coordination: Sense Organs
0-7167-9222-2 Urogenital System
0-7167-9288-5 Glossary of Vertebrate Anatomical Terms

LABORATORY SEPARATES BY ABRAMOFF/THOMSON

From *Laboratory Outlines in Biology VI*
1994, 526 pages, ISBN 0-7167-2633-5

0-7167-9081-5 Biologically Important Molecules: Proteins, Carbohydrates, Lipids, and Nucleic Acids
0-7167-9082-3 Light Microscopy
0-7167-9083-1 Cell Structure and Function
0-7167-9084-X Subcellular Structure and Function
0-7167-9085-8 Cellular Reproduction
0-7167-9086-6 Movement of Materials Through Plasma Membranes
0-7167-9087-4 Enzymes
0-7167-9088-2 Cellular Respiration
0-7167-9089-0 Photosynthesis
0-7167-9090-4 Mendelian Genetics
0-7167-9091-2 Chromosomal Basis of Heredity
0-7167-9092-0 Human Genetics
0-7167-9093-9 Expression of Gene Activity
0-7167-9094-7 Kingdom Monera
0-7167-9095-5 Kingdom Protista I: Algae and Slime Molds
0-7167-9096-3 Kingdom Protista II: Protozoa
0-7167-9097-1 Kingdom Fungi
0-7167-9098-X Kingdom Plantae: Division Bryophyta
0-7167-9099-8 Kingdom Plantae: The Vascular Plants
0-7167-9100-5 Plant Anatomy: Roots, Stems, and Leaves
0-7167-9101-3 Flowers and Fruits
0-7167-9117-X Transport and Coordination in Plants
0-7167-9118-8 Plant Growth and Development
0-7167-9119-6 Plant Development: Hormonal Regulation
0-7167-9102-1 Kingdom Animalia: Phyla Porifera, Cnidaria, and Ctenophora
0-7167-9120-X Kingdom Animalia: Acoelomates (Phylum Platyhelminthes) and Pseudocoelomates (Phyla Nemtoda and Rotifera)
0-7167-9103-X Kingdom Animalia: Phylum Mollusca
0-7167-9104-8 Kingdom Animalia: Phylum Annelida
0-7167-9105-6 Kingdom Animalia: Phylum Onychophora and Arthropoda
0-7167-9106-4 Kingdom Animalia: Phylum Echinodermata
0-7167-9121-8 Kingdom Animalia: Phyla Hermichordata and Chordata
0-7167-9107-2 Vertebrate Anatomy: External Anatomy, Skeleton, and Muscles
0-7167-9108-0 Vertebrate Anatomy: Digestive, Respiratory, Circulatory, and Urogenital Systems
0-7167-9109-9 Biological Coordination in Animals
0-7167-9110-2 Nervous System Physiology
0-7167-9111-0 Fertilization and Early Development of the Sea Urchin
0-7167-9112-9 Fertilization and Early Development of the Frog
0-7167-9113-7 Early Development of the Chick
0-7167-9114-5 Analysis of Surface Water Pollution by Microorganisms
0-7167-9115-3 Analysis of Solids in Water

(continued next page)

W.H. FREEMAN AND COMPANY
41 Madison Avenue, New York, NY 10010
Houndmills, Basingstoke, RG21 6XS, England

LABORATORY SEPARATES BY HELMS, KOSINSKI, AND CUMMINGS

From *Biology in the Laboratory*, Third Edition
1998, 720 pages, ISBN 0-7167-3146-0

0-7167-9303-2 Science—A Process and Appendixes I, II, and III
0-7167-9304-0 Observations and Measurements: Microscope
0-7167-9305-9 Observations and Measurements: Measuring Techniques
0-7167-9306-7 pH and Buffers
0-7167-9307-5 Using the Spectrophotometer
0-7167-9308-3 Organic Molecules
0-7167-9309-1 Prokaryotic Cells
0-7167-9310-5 Eukaryotic Cells
0-7167-9311-3 Osmosis and Diffusion
0-7167-9312-1 Mitosis
0-7167-9313-X Enzymes
0-7167-9314-8 Energetics, Fermentation, and Respiration
0-7167-9315-6 Photosynthesis
0-7167-9316-4 Meiosis: Independent Assortment and Segregation
0-7167-9317-2 Genes and Chromosomes: Chromosome Mapping
0-7167-9318-0 Human Genetic Traits
0-7167-9319-9 DNA Isolation and Protein Synthesis
0-7167-9320-2 DNA—The Genetic Material: Replication, Transcription, and Translation
0-7167-9321-0 Molecular Genetics: Recombinant DNA
0-7167-9322-9 Genetic Control of Development and Immune Defenses
0-7167-9323-7 The Genetic Basis of Evolution—Populations
0-7167-9324-5 The Genetic Basis of Evolution II—Diversity
0-7167-9325-3 Diversity—Kingdoms Monera and Protista
0-7167-9326-1 Diversity—Fungi and the Bryophytes
0-7167-9327-X Diversity—Vascular Land Plants
0-7167-9328-8 Diversity—Porifera, Cnidaria, and Wormlike Invertebrates
0-7167-9329-6 Diversity—Mollusks, Arthropods, and Echinoderms
0-7167-9330-X Diversity—Phylum Chordata
0-7167-9331-8 Plant Anatomy—Roots, Stems, and Leaves
0-7167-9332-6 Angiosperm Development—Fruits, Seeds, Meristems, and Secondary Growth
0-7167-9333-4 Water Movement and Mineral Nutrition in Plants
0-7167-9334-2 Plant Responses to Stimuli
0-7167-9335-0 Animal Tissues
0-7167-9336-9 Introduction to the Study of Anatomy, and the External Anatomy and Integument of Representative Vertebrates
0-7167-9337-7 The Anatomy of Representative Vertebrates: Behavioral Systems
0-7167-9338-5 The Anatomy of Representative Vertebrates: Digestive and Respiratory Systems
0-7167-9339-3 The Anatomy of Representative Vertebrates: Circulatory and Urogenital Systems
0-7167-9340-7 The Basics of Animal Form: Skin, Bones, and Muscles
0-7167-9341-5 The Physiology of Circulation
0-7167-9342-3 Gas Exchange and Respiratory Systems
0-7167-9343-1 The Digestive, Excretory, and Reproductive Systems
0-7167-9344-X Control—The Nervous System
0-7167-9345-8 Behavior
0-7167-9346-6 Communities and Ecosystems
0-7167-9347-4 Predator-Prey Relations
0-7167-9348-2 Productivity in Aquatic Ecosystems

LABORATORY SEPARATES BY WISCHNITZER

From *Atlas and Dissection Guide for Comparative Anatomy,* Fifth Edition
1993, 291 pages, ISBN 0-7167-2374-3

Anatomy of the Dogfish Shark
0-7167-9240-0 External Morphology
0-7167-9241-9 Skeletal System
0-7167-9242-7 Muscular System
0-7167-9243-5 Digestive and Resipiratory Systems
0-7167-9244-3 Circulatory System
0-7167-9245-1 Urogenital System
0-7167-9246-X Sense Organs
0-7167-9247-8 Nervous System

Anatomy of the Protochordates
0-7167-9239-7 Morphology

Anatomy of the Lamprey
0-7167-9263-X Morphology

Anatomy of the Mud Puppy *Necturus*
0-7167-9248-6 External Morphology
0-7167-9249-4 Skeletal System
0-7167-9250-8 Muscular System
0-7167-9251-6 Digestive and Respiratory Systems
0-7167-9252-4 Circulatory System
0-7167-9253-2 Urogenital System
0-7167-9254-0 Sense Organs and Nervous System

Anatomy of the Cat
0-7167-9255-9 External Morphology
0-7167-9256-7 Skeletal System
0-7167-9257-5 Muscular System
0-7167-9258-3 Digestive and Respiratory Systems
0-7167-9259-1 Circulatory System
0-7167-9260-5 Urogenital System
0-7167-9261-3 Endocrine System and Sense Organs
0-7167-9262-1 Nervous System

Warren F. Walker, Jr.
EMERITUS PROFESSOR, OBERLIN COLLEGE

ILLUSTRATED BY

Edna Indritz

Evanell Towne

Edward Hanson

W. H. FREEMAN AND COMPANY
New York

Acquisitions Editor: Patrick Shriner
Associate Editor: Debra Siegel
Assistant Editor: Melina LaMorticella
Project Editor: Sarah Kimnach, ESNE
Production Coordinator: Maura Studley
Administrative Assistant: Ceserina Pugliese
Cover and Text Design: John Svatek, ESNE
Manufacturing: Olympian Graphics

ISBN 0-7167-2636-X

Printed in the United States of America

First Printing 1988

Table of Contents

THESE BRIEF REMARKS are aimed at helping a teacher to decide quickly upon the material that is needed, demonstrations that might be set up, and parts of an exercise in which a student might need special help.

Exercise 1 External Anatomy and Skeleton

The student can make all of the dissections on a single frog; but, if the blood vessels are to be studied, injected specimens will be needed, and if both arteries and veins are to be examined, doubly injected specimens will be required. Although bullfrogs are more expensive, they are more satisfactory than leopard frogs because of their larger size. They should be used if all of the muscles described, the internal structure of the heart, and the eye are to be dissected. Most of the drawings are based on a bullfrog, but they are applicable to other frogs of the family Ranidae.

If a few live frogs are needed for demonstrations, they may be caught locally or, like preserved and injected frogs, obtained from biological supply houses. If the frogs are to be kept alive for only a few days, keep them in an aquarium in a cool room (about 50-60°F or 10-15°C). At these temperatures metabolism is sufficiently low that the frogs need not be fed. Have about two inches of water in the aquarium with *Rana pipiens,* but place rocks at one end so they can climb out. Water depth should be three or four inches for *Xenopus laevis,* and rocks are not necessary. Cover both aquaria with screening, for frogs can jump out. If live frogs are to be kept for long periods, more elaborate quarters will be needed, and the frogs should be fed. Disease control may also be a problem. The publication, *Amphibians, Guidelines for the breeding, care, and management of laboratory animals* (1974) provides all of the information you will need. This book is a report of the Subcommittee on Amphibian Standards, Committee on Standards, Institute of Laboratory Animal Resources, National Research Council. It is available from the Publishing Office, National Academy of Sciences, 2101 Constitution Ave., N. W., Washington, D. C. 20419.

The South African clawed frog, *Xenopus,* usually is sold living, and you should kill and preserve them if they are to be dissected. *Xenopus* is an interesting example of a highly aquatic species. You may wish to demonstrate some aspects of their anatomy even if they are not dissected by the whole class.

A. General External Features

A live frog and an aquarium should be available so that students can see how a frog jumps and swims. You may wish to have specimens of both *Rana* and *Xenopus.* Biological supply companies can provide instructions for keeping and raising *Xenopus* and can supply the special food that they will need.

B. Skin and Pigmentation

Students will need slides, cover glasses, and microscopes to prepare and study wet mounts of frog skin.

Preparing a demonstration of the color hues of a frog requires keeping one live frog in a white container beneath a light and another in a black container in the dark for four to five hours before class.

C. The Skeleton

Mounted skeletons of frogs are essential for this study, and it would be very helpful if isolated trunk vertebrae were provided. If possible, human skeletons should be available for comparisons.

Exercise 2 Muscles

No special preparations are suggested for this exercise.

Exercise 3 Digestive and Respiratory Systems

A. The Buccopharyngeal Cavity

Cut the lower jaw and the floor of the mouth off of a pithed frog, pin the frog down on its back, and place small pieces of cork or carbon particles on the roof of the buccopharyngeal cavity to demonstrate ciliary action.

C-D. General Visceral Organs, Digestive Tract

Another pithed frog should be sacrificed so that students can see the viscera in their natural condition. By carefully inserting a small glass tube into the glottis, the entrance of a vocal sac in a male, and the cloaca, it is possible to inflate the lungs, a vocal sac, and the urinary bladder. This will give students a better appreciation of the size of these organs in life.

E. The Respiratory System

One live frog should be available so that the students may observe the breathing movements.

Exercise 4 Circulatory System

C. The Venous and Lymphatic Systems

At least one frog must be pithed so that capillaries and the caudal lymph hearts can be demonstrated. If the same frog is to be used for both, demonstrate the lymph hearts first. They can be found by reflecting a flap of skin from the back over the urostyle and pelvic girdle. It is unlikely that you will be able to show more than their pulsation, because the lymph hearts are so thin-walled that it is difficult to dissect them free from surrounding tissues without destroying them.

To show the capillaries, wrap the frog in a wet cloth with a hind leg protruding, stretch the web, and pin the foot over a hole in a cork sheet or in a piece of wood. Keep the web wet by adding drops of Ringer's solution. Observe with the low power of a microscope. Remember that because structures are greatly magnified, the apparent rate of movement of blood in the capillaries is also magnified.

D. Heart

Either the frog used for the procedures above, or another pithed frog should be cut open so that the students can observe the beating of the heart. Frequently add a few drops of Ringer's solution to the heart. The effect of various drugs, such as adrenaline, on the rate of heart beat can also be demonstrated by adding a few drops of such substances directly on the heart.

If the internal structure of the heart is to be studied, it would be well to prepare one demonstration dissection of the heart of a specimen that has not been injected. It is sometimes difficult to remove all of the latex from an injected heart without injuring the valves.

Exercise 5 Urogenital System

A. The Excretory System

If students did not see a demonstration of an inflated urinary bladder during Exercise 3, one should be prepared at this time.

Exercise 6 Nervous System

B. The Sympathetic Cord

It may be necessary to use dissecting microscopes to find all of the communicating branches and to see the sympathetic ganglia clearly.

C. The Spinal Cord and the Origin of the Spinal Nerves

Dissection of the spinal cord and brain can be facilitated by putting the frog in a weak (3 to 6%) solution of nitric acid for several days. This will decalcify the bone. Dissecting microscopes are essential if the roots of a spinal nerve are to be studied.

D. The Brain

Some of the activities of the nervous system can be demonstrated by the destruction of certain parts of the central nervous system. There are directions for performing such experiments in most biology manuals.

If flight from predators and feeding behavior are to be demonstrated, you will probably have more success with a toad than with a frog. Meal worms are a choice food. Simply place a meal worm 8 to 10 centimeters away from the frog, or dangle it on a thread close to ground level. Prey must be moving to stimulate the feeding response.

E. The Cranial Nerves

Because it is unlikely that students can find all of the cranial nerves on their own specimens, a demonstration dissection should be provided.

Exercise 7 Sense Organs

B. The Eye

The structure of the eyeball cannot be seen clearly unless bullfrogs are used. Even for bullfrogs dissecting scopes and sharp scissors with fine points will be needed.

C. The Ear

If the skull has been thoroughly decalcified, it is possible to pick away bone carefully enough to expose a membranous labyrinth. It would be desirable to have a labyrinth on demonstration.

THIS SET OF EXERCISES on the *Dissection of the Frog* is a reprint, with corrections and updating, of the second edition, originally published in 1980. The second edition of *Dissection of the Frog,* like the first, has been prepared to meet the needs of introductory college and advanced high school students for a set of exercises that describe the frog's gross anatomy with reasonable thoroughness. Being a good example of a vertebrate, the frog is one of the most widely used animals in vertebrate physiology and in other experimental work. It is clearly an animal with which biology students should be familiar. It is hoped also that these exercises will enable certain advanced students, whose introductory work did not include this interesting animal, to study the frog on their own, or as a special project for a course in which a knowledge of this vertebrate is relevant. Certain exercises may also be assigned as gross anatomical background by instructors using the frog in upper division physiology, embryology, or histology classes.

Major changes in this edition have been the addition of material on the South African clawed frog, *Xenopus laevis,* and a glossary of anatomical terms. Although *Xenopus* is not commonly dissected in introductory classes, it is frequently used in embryonic and experimental studies because it is easy to breed and raise in the laboratory and because it grows rapidly. Many students may wish to become familiar with this interesting example of an aquatic species that is neotenic and retains many larval features throughout its life. As our native frogs become rarer and more expensive, *Xenopus* doubtless will be more widely used.

The glossary provides students with a basic vocabulary of anatomical terms, which should be the easier for students to remember because the classic derivations are also given. If students conscientiously refer to these derivations, they will become familiar with common roots that are used repeatedly in anatomical terminology.

Some sections of the exercises have been modified to clarify dissection procedures, some have been updated to include new discoveries, and some have been expanded to explain more fully the functions of the parts being studied.

Each exercise is based on a natural unit of material, usually one or two organ systems. Gross anatomy is emphasized, but considerable stress is put on the functions of the various parts, and on their functional interrelations with other parts. Gross structure cannot be understood without perspectives like these.

Detailed directions for dissection are included to enable a student to proceed with minimum help from the instructor, and to encourage the student to take more than a superficial look at the material. The careful observations necessary to verify the descriptions should help a student to appreciate both the potentials and the limitations in the descriptive aspects of science. And careful observations, of course, are necessary for the student's evaluations of the generalizations that are derived from an anatomical study. In some of the exercises, demonstrations of freshly killed, or pithed, frogs are suggested, because certain structures can be seen particularly well in fresh material.

The drawings have been selected to help a student find the essential parts, and to serve as a permanent reference to the structures observed in dissection, after the specimen has been disposed of. Many of the drawings are of the type used in atlases, showing structures in their relationships to surrounding parts; these not only make it easier to identify the structures, but emphasize the unity of the organism. Some students may wish to make additional sketches of their own. If time permits, this is to be encouraged, because a very careful dissection and a close observation are required before an organ can be drawn accurately.

In these exercises, technical terms are printed in **bold face** type when first used. Anatomy is plagued with a surplus of synonymous terms. Where two terms are in common usage, both are given, but the favored one is mentioned first. For the most part, favored terms are anglicized versions of those recommended by the *Nomina Anatomica Veterinaria.* This code, which applies in so far as possible the human terms of the *Nomina Anatomica* to quadrupeds, is becoming the standard for mammals other than human beings. The two codes differ primarily with respect to modifying adjectives for direction; thus, the human "inferior vena cava" is referred to as the "caudal vena cava." Although most of the *Nomina* terms can be applied to homologous structures in lower vertebrates, it must be remembered that the frog and lower vertebrates have some unique structures not present in mammals.

Many people have encouraged and helped me in preparing both editions of this manual. I continue to be indebted to Miss Edna Indritz, Miss Evanell Towne, and Mr. Edward Hanson for the great care and artistry with which they prepared final drawings for the first edition from my sketches. Miss Indritz has done the same for new illustrations for this edition. Miss P. Anne Smith, an Oberlin student, helped me prepare sketches for the new figures. I also wish to thank the numerous users of the first edition who have taken the time and trouble to send comments to the publisher. Their encouragement has been gratifying, and their criticisms have formed the basis for many of the changes in this edition. It is my hope that users of these exercises will continue to call my attention to any errors or parts in need of revision. My wife, Hortense, has been most helpful in proofreading. Finally, I am very grateful to Kevin Gleason, who skillfully edited the second edition, and to the production department of W. H. Freeman and Company for their support and help.

This reprint has been completely reset with corrections and updating, and was carefully edited by Sarah Kimnach of ESNE (Editorial Services of New England, Inc.). I am most grateful for her help. The Glossary of Vertebrate Anatomical Terms, which is also used in *Anatomy and Dissection of the Fetal Pig* and *Anatomy and Dissection of the Rat,* has been expanded and thoroughly revised in collaboration with Professor Dominique G. Homberger of Louisiana State University, Baton Rouge. I am very much indebted to her for her help.

Warren F. Walker, Jr.
Ossipee, New Hampshire, 1997

I F THIS IS THE FIRST ANIMAL that you have dissected, you should read these remarks concerning procedures and terms of direction carefully before you begin. At the outset, a few incisions will have to be made with scalpel or scissors to open the specimen, but thereafter, most of the dissecting should be done with a pair of fine forceps. Dissecting consists of carefully separating organs and picking away surrounding connective tissue in order to expose the organs clearly. Do not cut out any organs unless specifically directed to do so, but always pick away enough of the surrounding tissue to see all of an organ clearly. Do not be content with just a glimpse of it. A good dissection should reveal all of the organs clearly enough so that another person would have no difficulty examining the specimen and seeing the essential relationships of the organs and their connections with other organs.

The terms *right* and *left* always refer to the *specimen's* right and left. Depending on how the specimen is oriented, this may or may not correspond with your right and left. *Lateral* refers to the side of the body or organ in question; *medial*, to the center. Other terms for direction differ somewhat for a quadruped and for a human being who stands erect (see the accompanying figure). In a human, *superior* refers to the upper, or head end, of the body; *inferior*, to the lower parts of the body. Our belly surface is *anterior* and our back is *posterior*. In a quadruped, the underside of the body is *ventral* and the back is *dorsal*. *Anterior* and *posterior* are acceptable terms of direction for some structures in a quadruped head because this part of the body corresponds in its orientation to the human head. However, because structures in other parts of the quadruped body have a different orientation, there is a tendency to drop the terms anterior and posterior in most parts of the quadruped and to replace them with *cranial*, for a direction toward the head, and *caudal*, for a direction toward the tail.

Certain structures may be described as *superficial*, by which is meant that they lie close to the surface or above some other structure. A *deep* structure lies beneath another and farther away from the body surface.

The *distal end* of a structure is the end farthest away from some point of reference, usually the origin of the structure or the midventral line of the body; the *proximal end* is the end that is nearest to the point of reference.

A *sagittal section* is a median section in the longitudinal plane of the body passing from dorsal to ventral. A *frontal section* is also a median longitudinal section, but passes from right to left. A *transverse section* crosses the longitudinal axis of the body at right angles.

At the end of a period always clean up your work area and put your specimen away in the container provided.

References

The list of references below will be of use to those who wish to pursue various aspects of anuran biology. The list is not intended to be exhaustive; it is an introduction to the extensive literature that is available, and only those references that are of particular value are cited. Some of the references are very old, but frogs have been studied for many years, and certain of the classic work has not been surpassed. Unfortunately, some items are out of print, but these can be found in most large libraries.

Behler, J. L. and F. W. King. *The Audobon Society Field Guide to North American Reptiles and Amphibians.* New York: Alfred A. Knopf, 1979.

An authoritative and easy to use guide for the identification of frogs and other amphibians and reptiles.

Biological Abstracts. Philadelphia, 1926 to date.

An essential bibliographic tool for those who wish to check the primary research literature. Two issues summarizing the major biological research in the world are published monthly.

Cochran, D. C. *Living Amphibians of the World.* Garden City: Doubleday, 1961.

An excellent and superbly illustrated account of the natural history of amphibians. A section is devoted to each family.

Conant, R., and Collins, J. T. *Reptiles and Amphibians of Eastern and Central North America,* 3rd ed. Boston: Houghton Mifflin Company, 1991.

A Peterson Field Guide.

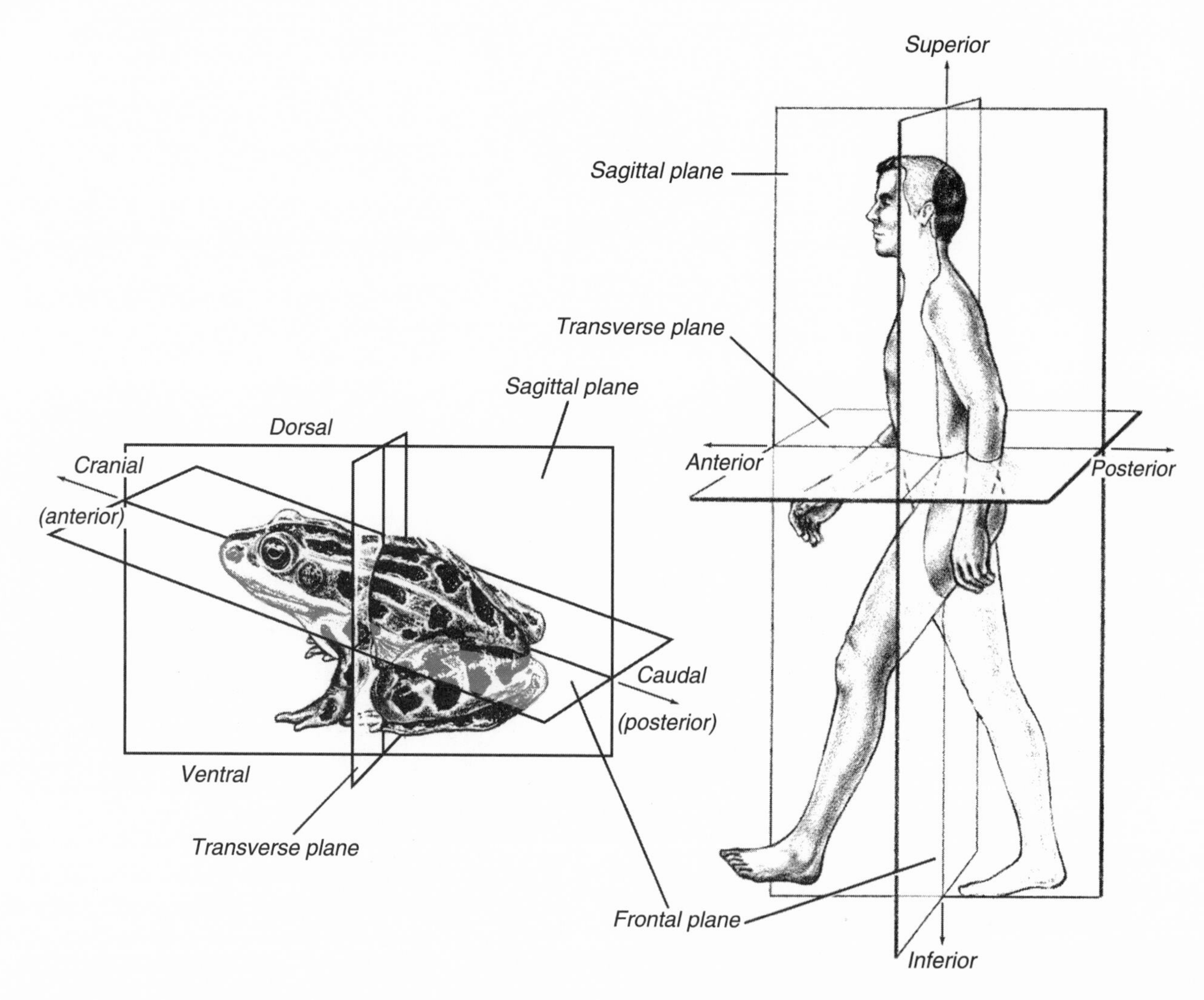

de Jongh, H. J., and C. Gans. *On the mechanism of respiration in the bullfrog,* Rana catesbeiana: *A reassessment,* Journal of Morphology, 127: 259-290, 1969.

A definitive study of breathing mechanisms.

Deuchar, E. *Xenopus, the South African Clawed Frog.* New York: John Wiley and Sons, 1975.

An account of the biology of *Xenopus* with emphasis on its reproductive biology and development.

Dorit, R. L., Walker, W. F., Jr., and Barnes, R. D. *Zoology.* Philadelphia: Saunders College Publishing, 1991.

Includes a great deal about frog biology.

Duellman, William E., and Trueb, Linda. *Biology of the Amphibia.* New York: McGraw-Hill Book Company, 1986.

The contemporary successor to Noble's classic text. It deals with all aspects of amphibian biology citing numerous references to the research literature.

Dunlap, D. G. *Comparative myology of the pelvic appendage in the Salientia,* Journal of Morphology 106: 1-76, 1960.

A study of the morphology of a region often dissected by students of elementary zoology.

Ecker, A. *The Anatomy of the Frog.* Oxford: Clarendon Press, 1889.

An English translation by S. Haslam of Ecker's German book on frog anatomy. It is perhaps more available than Gaupp's revision and expansion of Ecker's sudy, but it is not as thorough.

Ewert, J. P. *The neuronal basis of visually guided behavior,* Scientific American 230, No. 3: 34-42, 1974.

A summary of research on prey-catching and avoidance behavior in the European toad.

Feder, M. E., and Burggren, W. W. (eds.) *Environmental Physiology of the Amphibians.* Chicago: University of Chicago Press, 1992.

An excellent source book for the serious student.

Frazer, J. F. D. *Amphibians.* New York: Springer-Verlag, 1973.

A concise account of many aspects of amphibian biology written for advanced high school students and college undergraduates.

Frewein, J., Habel, R. E., and Sack, W. O. *Nomina Anatomica Veternaria,* 4th ed. Cornell University, Ithaca, NY: World Association of Veterinary Anatomist, 1984.

A list of standardized Latin names for organs and anatomical structures for quadruped mammals. Most are applicable to nonmammalian tetrapods.

Gaupp, E. *Anatomie des Frosches.* Braunschweig: Friedrich Vieweg und Sohn, 1891-1904.

This is an expanded edition of an earlier treatise by Ecker and Wiedersheim. This is the most thorough study of frog anatomy and histology in existence. It is still the point of departure for research in anuran morphology.

Gilbert, S. G. *Pictorial Anatomy of the Frog.* Seattle: University of Washington Press, 1965.

An excellent atlas of frog anatomy

Hetherington, T. E., Jaslow, A. P., and Limbered, R. E. *Comparative morphology of the amphibian opercular system. I. General design features and functional interpretation.* Journal of Morphology 190: 42-61, 1986.

Holmes, E. B. *A reconsideration of the phylogeny of the tetrapod heart.* Journal of Morphology 147: 209-228, 1975.

Holmes, S. J. *Biology of the Frog.* 4th ed., New York: Macmillan, 1934.

In the 1930s and '40s, this was a widely used textbook of frog biology. It is still the most available summary of Gaupp's classic work.

Johansen, K., and Hanson, D., *Functional anatomy of the hearts of lungfishes and amphibians.* American Zoologist 8: 191-210, 1968.

Moore, J. A. (ed.) *Physiology of the Amphibia.* New York: Academic Press, 1964, 1974 and 1976.

A valuable three-volume source book covering most aspects of amphibian physiology.

Noble, G. K. *Biology of the Amphibia.* New York: McGraw-Hill, 1931; reprint, New York: Dover Publications, 1954.

A standard source book dealing with the evolution, anatomy, physiology, and ecology of amphibians.

Rugh, R. *The Frog, Its Reproduction and Development.* Philadelphia: Blakiston, 1951.

A thorough account of the embryology of the frog through its metamorphosis.

Stebbins, R. C. *A Field Guide to Western Reptiles and Amphibians.* Boston: Houghton Mifflin Company, 1966.

Tyler, M. J. *Frogs.* Sydney: Collins, 1976.

A fascinating account of frogs, emphasizing their behavior and natural history.

Walker, W. F., Jr., and Liem, K. F. *Functional Anatomy of the Vertebrates: An Evolutionary Perspective.* Philadelphia: Saunders College Publishing, 1994.

A comparative anatomy text analyzing structure in functional terms.

Wright, A. H., and A. A. Wright. *Handbook of Frogs and Toads of the United States and Canada.* 3rd ed., Ithaca: Comstock, 1949.

This is the standard reference book on the systematics and the natural history of the frogs of North America.

External Anatomy and Skeleton

FROGS ARE OFTEN STUDIED in general biology and zoology courses because they are inexpensive and easy to obtain and dissect, and their structure shows clearly the major features of the organ systems characteristic of the backboned animals, or vertebrates. Vertebrates are of particular interest to us because we too are members of the subphylum Vertebrata of the phylum Chordata. Other chordate subphyla include a few soft-bodied marine animals such as sea squirts and amphioxus. Being members of the class Amphibia, frogs also have characteristics that are between those of fishes and those of terrestrial vertebrates, the reptiles, birds, and mammals.

Contemporary amphibians are grouped into three orders: frogs and toads (order Anura), salamanders (order Urodela), and legless, worm-shaped caecilians of the tropics (order Gymnophiona). All reproduce in the water or in a very moist place on land. Most go through an aquatic larvae stage (the larva is commonly called the tadpole). This period of development usually is followed by a metamorphosis to a terrestrial adult. Adult amphibians have only a rudimentary ability to conserve body water; hence they must live in damp habitats on the land, and many often return to the water. As a result of their phylogenetic position between fishes and reptiles and their double mode of life in water and on land, amphibians have a mixture of aquatic and terrestrial attributes.

There are many different species of frogs. Most of the North American ones belong to the family Ranidae and are exemplified by the leopard frog or meadow frog, *Rana pipiens,* and by the bullfrog, *Rana catesbeiana.* Although this dissection guide is based primarily on these two species, the clawed frog of Central and South Africa, *Xenopus laevis,* has been included as well because it is easy to breed and raise in captivity and it is widely used in experimental work. *Xenopus* belongs to the family Pipidae and differs from ranid frogs in being primarily an aquatic species. It leaves ponds only on damp nights, when it may travel overland to other bodies of water. During hot, dry weather, *Xenopus* can burrow into the mud at the bottom of a drying pond and aestivate. It has adapted to aquatic life chiefly by retaining certain characteristics of the aquatic larva throughout its life. This has been accomplished by slowing down the embryonic development of many organs relative to sexual development so that the animal becomes sexually mature without fully metamorphosing. This condition is called **neoteny.**

A. GENERAL EXTERNAL FEATURES

The body of a frog can be divided into a **head,** which extends caudally to the shoulder region, and a **trunk** (Figs. 1-1 and 1-2). Fishes, which were ancestral to amphibians, also have a powerful tail used in swimming. Aquatic tadpoles retain such a tail, but it is lost in the adult frog, which moves instead by means of its powerful hind legs. The much smaller front legs are used primarily to keep the front of the body raised from the ground or pond bottom. *Xenopus* also uses its front legs to push food into its mouth, for it lacks the tongue that ranid frogs use to catch food. Observe how a live frog jumps and swims. Notice that a distinct neck is absent. This is a retention of a characteristic of fishes, for which an independent motion of the head and trunk would be disadvantageous during swimming.

The combined orifice of the digestive and urogenital tracts, the cloacal aperture, is at the caudal end of the trunk just dorsal to the junction of the hind legs. The cloacal aperture of female Xenopus is flanked by a pair of papillae known as cloacal lips. This is one example of a structural difference between the sexes that is called sexual dimorphism. The anus of human beings and most other mammals, because it is the opening only of the digestive tract, is only partly comparable to the cloacal aperture.

General External Features

A large **mouth,** a pair of external nostrils, or **nares** (singular **naris**), and the **eyes** will be recognized on the head. In ranid frogs, the upper **eyelid** is a simple fold of skin, but the lower lid is a transparent membrane that can be drawn across the surface of the eyeball. Eyelids are absent or poorly developed in most aquatic vertebrates, and those of *Xenopus* are much smaller. The disc-shaped area caudal to each eye in ranid frogs is the ear drum, or **tympanic membrane.** Although it is the same size in male and female leopard frogs, it is considerably larger in male bullfrogs than in females (Fig. 1-3). The tympanic membrane is an adaptation to detect airborne sound waves and does not appear until metamorphosis. As you might expect, it is absent in *Xenopus.* Look carefully at the top of the head of *Rana* and, between the eyes, you may see a small, light **frontal organ,** or **brow spot,** about the diameter of a pin. This is a remnant of a median, light-sensitive eye that characterized primitive groups of fishes and amphibians. Look carefully at the head of *Xenopus* and observe the row of small, whitish dashes that encircle each eye, ramify over the top of the head and jaw margins, and continue caudally along each side of the flank. They constitute the **lateral line system,** an aquatic sensory system consisting of numerous microscopic receptors that detect low-frequency vibrations and other movements in the aquatic environment. The lateral line system has been called a system of "distant touch," for it enables the animal to detect the presence of other organisms or disturbances in the water. Such a system is found in all fishes and larval amphibians, but it is lost at metamorphosis in ranid frogs.

The legs of a frog have the same parts as our own. In the front leg, or pectoral appendage, notice the upper arm, the elbow, the forearm, the wrist joint, and the hand. There are only four fingers. Although the finger closest to

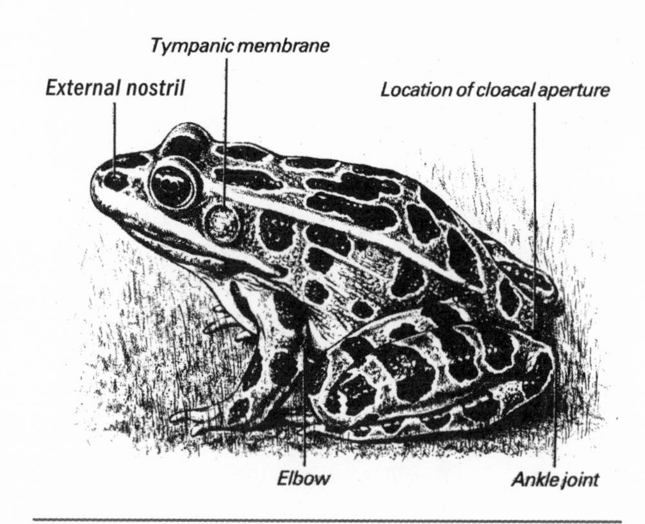

FIGURE 1-1
Lateral view of a leopard frog, *Rana pipiens.*

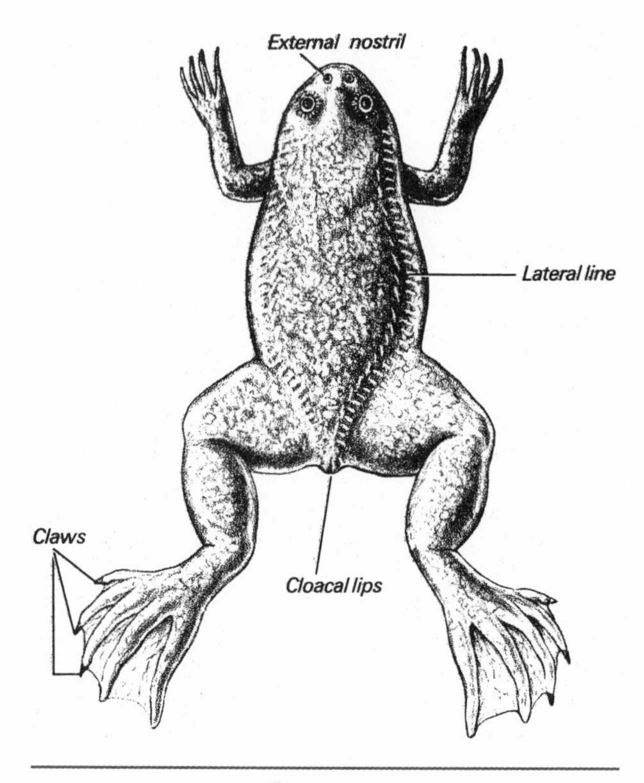

FIGURE 1-2
Dorsal view of a female African clawed frog, *Xenopus laevis.*

FIGURE 1-3
Lateral view of the head of a male (left) and female (right) bullfrog, *Rana catesbeiana,* showing the sexual dimorphism of the tympanic membrane.

the body is probably comparable to our second digit, it is often called the thumb because it is stouter than the others. During the breeding season, the thumb is particularly stout and darkly pigmented in the males of many species of frog—another example of sexual dimorphism. In the hind limb, or **pelvic appendage,** observe the thigh, the knee, the shank, the ankle joint, and the very long foot. Two elongated ankle bones lie within the proximal part of the foot, and the distal part bears five toes with a conspicuous web between them. The first toe is the smallest and most medial one, and it is comparable to our great toe. Technically it is called the **hallux,** and the small spur at its base in many anurans is the **prehallux.** The prehallux is much larger in toads, which use their hind feet in burrowing. The first three toes of *Xenopus* bear claws that help the frog grip the bottom. Its third, fourth, and fifth toes are exceptionally long and support a very large web.

B. SKIN AND PIGMENTATION

The skin of a vertebrate is a protective layer separating the internal and external environments. The skin, however, does not isolate the animal from the external environment; numerous receptors are found in it, and some exchanges of material occur across it. Frogs' skin is exceptionally thin as well as vascular. Secretions of numerous microscopic mucous glands keep the skin moist and help to protect the animal from ectoparasites. A thin, moist, vascular skin is an effective respiratory membrane, and a great deal of gas exchange between the blood and the environment takes place through the skin. At normal temperatures on land, most oxygen enters the body through the lungs, but most of the carbon dioxide leaves through the skin. Exchange of gasses through the skin is referred to as **cutaneous respiration.** However, the characteristics of the skin also permit considerable exchange of water across it—a loss of body water when on land, a gain of water when in an aquatic environment. Most of the water that the frog needs is absorbed by osmosis through the skin.

Skin and Pigmentation

Remove a small piece of skin from a lightly pigmented portion of the body, such as the throat, and make a wet mount. The small, branching, dark spots that you see with a microscope are the pigment cells, or **chromatophores.** These particular cells contain granules of a dark pigment known as **melanin.** The granules can migrate out the processes of the chromatophores, making the frog darker, or they can concentrate near the center of the cells, giving the frog a lighter color (Fig. 1-4). This difference in hues can be seen if one frog has been kept for several hours on a light background and another on a dark background. Other pigment cells contain a yellow pigment, and others contain refractive crystals of guanine, which disperse light to give a blue effect. The combination of yellow and blue causes the green color.

The color pattern of the frog is said to be **cryptic,** because it conceals the frog in its natural habitat. The greenish color of the back of a ranid frog blends with the natural background, and the dark spots and blotches form a disruptive pattern, which tends to obscure the shape of the animal. The dark back and the light belly—dark where the body is highlighted by natural illumination and light where there is normally a shadow—reduces the appearance of solidity, or mass, when the frog is viewed from the side.

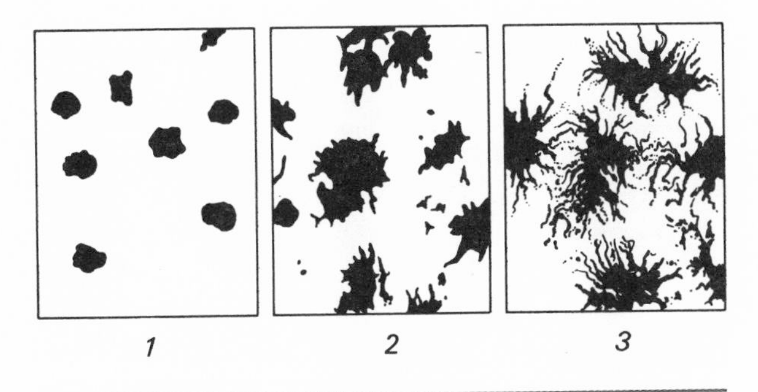

FIGURE 1-4
Three stages in the dispersal of pigment in the chromatophores of a frog. In stage 1, the pigment is concentrated near the center of the chromatophore, and the animal is light in color. Stage 2 is intermediate. In stage 3, pigment is fully dispersed and the animal appears dark.

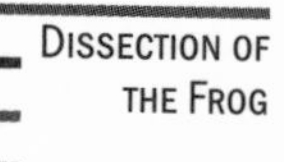

The Skeleton

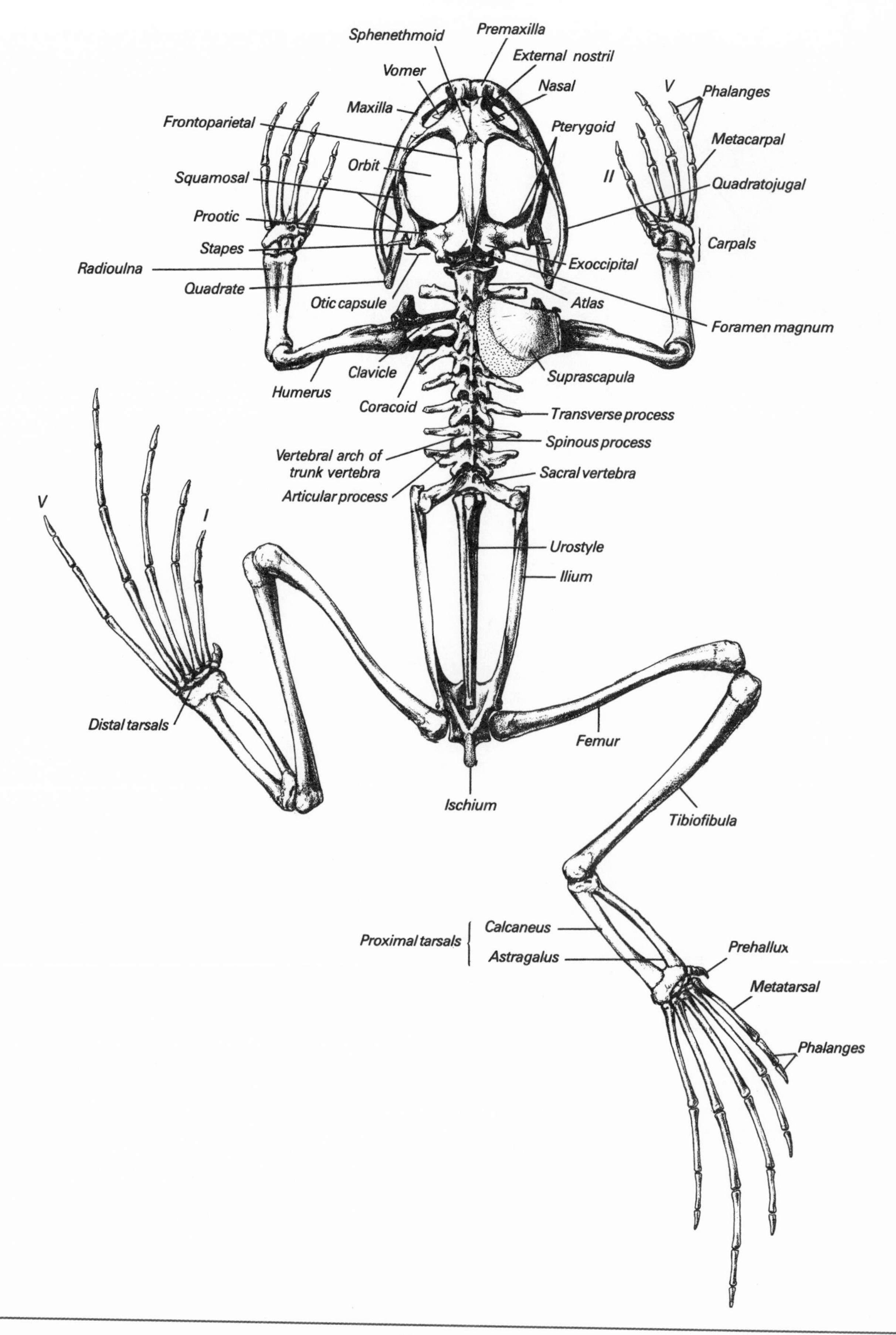

Figure 1-5
Dorsal view of the skeleton of a ranid frog.
Major cartilaginous areas in this and subsequent drawings of the skeleton have been stippled. [In part after Gaupp.]

C. THE SKELETON

Examine a mounted frog, and, if possible, compare it with a human skeleton to see both the similarities and the differences. The skeleton of any vertebrate is an internal framework of bone and cartilage that supports the body, forms lever arms that transfer the forces of muscle action to points of application, and protects such delicate organs as the brain and spinal cord. It can be divided into a **somatic skeleton,** located in the body wall and appendages, and a less conspicuous **visceral skeleton,** located chiefly in the wall of the cranial part of the digestive tract. The somatic skeleton is subdivided into an **axial skeleton,** which is in the longitudinal axis of the body (skull, vertebrae, ribs and breastbone), and the **appendicular skeleton,** which supports the appendages.

1. The Skull

The skull is the cranial part of the axial skeleton that houses the brain and special sense organs and that forms the jaws (Figs. 1-5 and 1-6). It can be divided into two parts: one houses the brain and inner ear (the **cranial region**); the other forms the jaws and helps to enclose the nose, the eye, and part of the ear (the **facial region**). The major bones and cartilages in these regions in *Rana* can be identified from the figures.

In the frog, and in lower vertebrates generally, only the narrow, longitudinal part of the skull along the midline constitutes the brain case, or **cranium.** Notice how the frog's cranium differs in this respect from the human cranium (Fig. 1-7). The large opening at the very caudal end of the cranium, through which the spinal cord enters the cranial cavity, is the **foramen magnum.** The caudal expansion of the cranial region on each side, which extends laterally toward the region that holds the tympanic membrane, is the **otic capsule.** It houses the inner ear, that part of the ear containing the receptive cells for equilibrium and sound (Exercise 7). Sound waves are transmitted from the external ear, which consists of the tympanic membrane held by a ring-shaped **tympanic cartilage,** across the **tympanic cavity** (the cavity of the middle ear) to the inner ear by a slender rod, the **stapes,** or **columella,** which can be seen in some specimens.

The **upper jaw** on each side is formed by two toothbearing bones, the **premaxilla** and **maxilla,** and a **quadrate** cartilage. The caudal ends of the quadrate cartilage and **mandibular cartilage,** which makes up much of the lower jaw, or **mandible,** form the jaw joint. Notice that the bones of the lower jaw do not have teeth. Large **orbits,** in which the eyes are lodged, lie between the upper jaws and brain case. The **external nostrils** lie on the dorsal surface of the skull just behind the front of the upper jaws. Each leads into a nasal passage that opens into the roof of the mouth by **internal nostrils,** or **choanae.** Tooth-bearing **vomer bones** form the roof of the mouth medial to the internal nostrils in ranid frogs. *Xenopus* lacks these teeth.

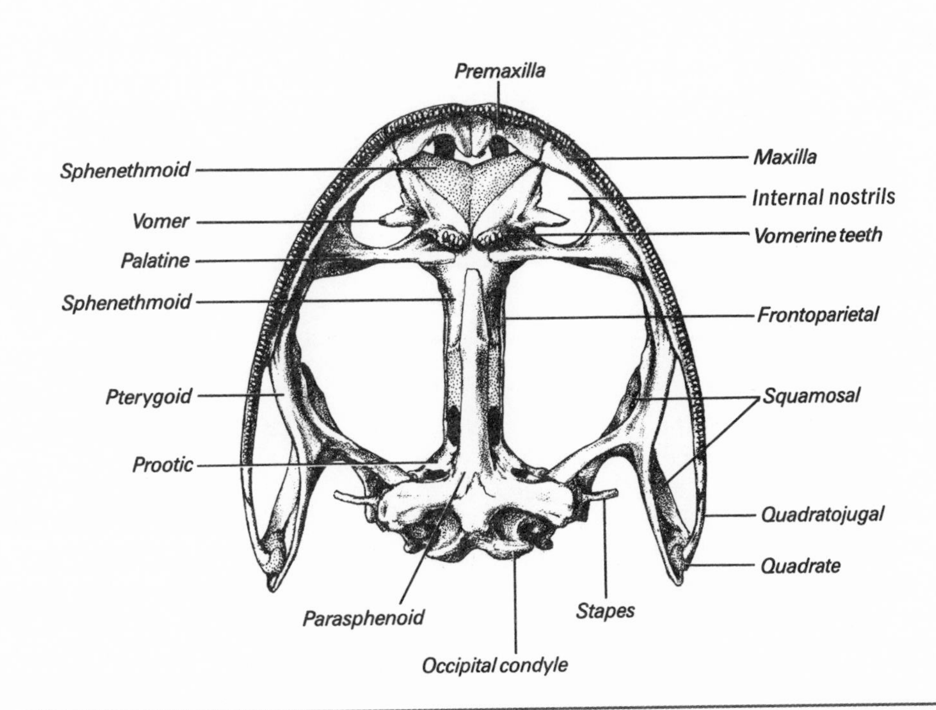

FIGURE 1-6
Ventral view of the skull of *Rana*. [After Gaupp.]

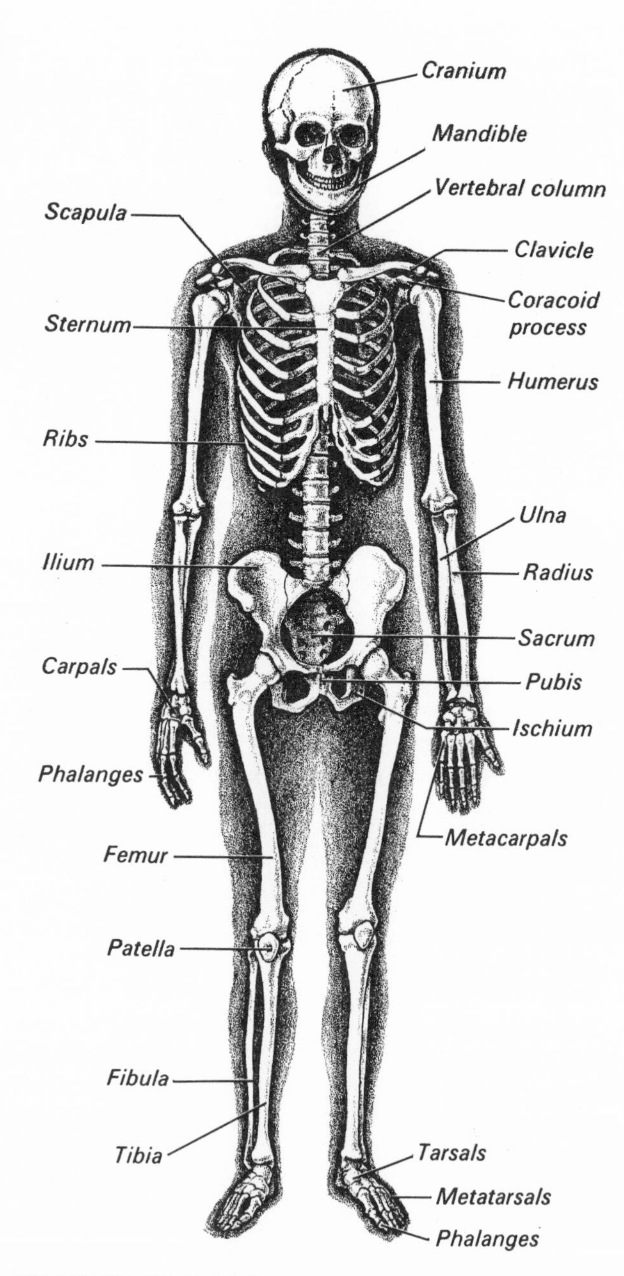

FIGURE 1-7
Anterior view of a human skeleton.

The Skeleton

2. The Visceral Skeleton

Examine the visceral skeleton before continuing with the axial skeleton, because it is located in the head region along with the skull. In very primitive fishes, the visceral skeleton consists of a series of cartilaginous or bony arches embedded in the lateral wall and floor of the pharynx, one at the base of each gill. With the loss of gills in terrestrial vertebrates, much of the visceral skeleton was also lost. Of the part of the visceral skeleton that remains, part has contributed to the skull (the quadrate cartilage, mandibular cartilage, and the stapes are of visceral origin), but most form the **hyoid apparatus** (Fig. 1-8). Most laboratory skeletons have the hyoid apparatus mounted separately beside the skeleton, but in the living vertebrate, it is a plate of cartilage on the floor of the pharynx at the base of the tongue. Arches of cartilage and bone extend from it around the pharynx to attach to the skull. The hyoid apparatus forms a sling that supports the tongue. A few parts of the visceral skeleton have become lodged in the walls of the glottis, the entrance to the larynx (Exercise 3), but these are very small and hard to find.

3. The Vertebrae

The short, compact, and strong vertebral column of the frog has evolved as an aid in the animal's jumping. It contains four types of vertebrae. The first vertebra, the **cervical vertebra,** or **atlas,** is modified, and it articulates with the skull. Notice that it lacks the conspicuous, laterally projecting pair of **transverse processes** that characterize the second type of vertebra—the seven **trunk vertebrae.** The ninth, or **sacral vertebra** of *Rana* has a particularly stout pair of transverse processes that articulate with the pelvic girdle, the arch of bone supporting the hind legs. Notice the greater number of sacral vertebrae in a human skeleton (Fig. 1-7). The fourth type of vertebra, the long prong of bone constituting the caudal part of the vertebral column, is the **urostyle.** It is formed by the fusion of two embryonic tail, or **caudal vertebrae.** The sacral vertebra and urostyle are fused in *Xenopus,* and the transverse process of the sacral vertebra is exceptionally broad and wing-shaped.

Examine an isolated trunk vertebra. Its ventral portion is a small block of bone known as the **vertebral body,** or **centrum.** Being concave at one end and convex on the other, the vertebral bodies interlock securely. In ranid frogs the concavity is **procoelous;** it is located on the cranial surface. In *Xenopus* it is **opisthocoelous,** that is, on the

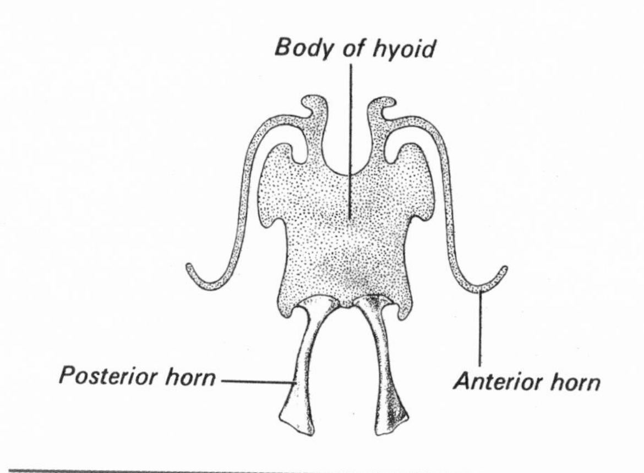

FIGURE 1-8
Dorsal view of the hyoid apparatus of *Rana.* [After Gaupp.]

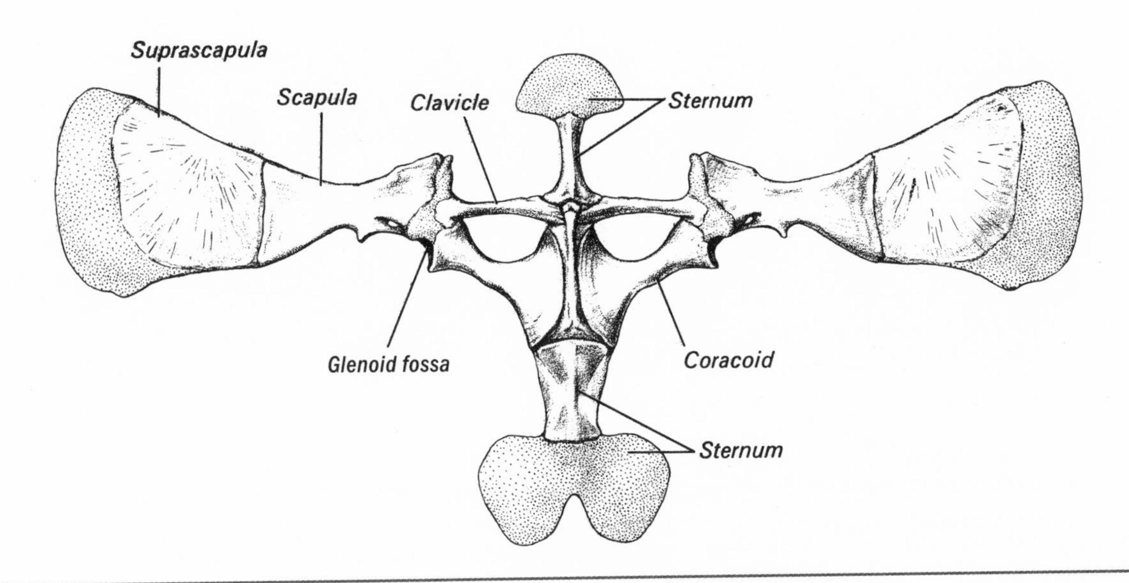

FIGURE 1-9
Ventral view of the sternum and pectoral girdle of *Rana*.
The scapula extends dorsally from the glenoid fossa, but it has been turned ventrally. [After Gaupp.]

caudal surface. An arch of bone, the **vertebral arch,** lies dorsal to the vertebral body and encases the spinal cord. It bears the **transverse processes** and also a small, dorsally projecting **spinous process.** Muscles and ligaments attach to these processes. Observe in the articulated skeleton how the lateral edges of the vertebral arches of successive vertebrae overlap; that is, the caudolateral corner of one vertebral arch passes beneath the craniolateral corner of the arch of the vertebra behind it. These overlapping regions are the **articular processes.**

In most vertebrates, ribs attach to many of the vertebrae, but these are absent in the adults of all but a few very primitive species of frogs. In some species of frogs, short ribs are present during the larval period, but they fuse with, and contribute to the transverse processes.

4. The Appendicular Skeleton and the Sternum

The pectoral appendages are supported by an arch of bone, the **pectoral girdle** (Fig. 1-9), but unlike the pelvic girdle, it does not have bony processes that connect it with the rest of the skeleton. Muscles transfer body weight from the axis of the frog to the girdle. On each side, the girdle contains a **scapula,** which extends dorsally from the socket for the arm (**glenoid fossa**) A **suprascapula,** which is only partly ossified, arches over the back from the top of the scapula. A cranial **clavicle** and a caudal **coracoid** extend toward the midventral line from the glenoid fossa. Those from the opposite sides meet to form the main struts that keep the glenoid fossae apart. The breastbone, or **sternum,** consists of the bone and cartilage extending cranially from the clavicles and caudally from the coracoids.

The Skeleton

The **humerus** is the bone of the upper arm (Fig. 1-5). Most terrestrial vertebrates have a radius and an ulna in the forearm (Fig. 1-7), as can be seen in a human skeleton. In the frog, radius and ulna have fused together to form a **radioulna.** The frog forearm has no need for rotation, which two bones permit, but the frog does require a firm strut to hold its body off the ground, and to act as a shock absorber when it lands after a jump. Several small **carpal bones** are in the wrist region; **metacarpals** occupy the palm of the hand; and **phalanges** form the parts of the digits extending beyond the palm. Note the number in each finger. A small spur of bone, which may be a vestige of the first metacarpal, can also be seen.

Each side of the **pelvic girdle** has a socket, the **acetabulum,** for the articulation of the hind leg (Fig. 1-10). The long bone extending from the acetabulum to the sacrum is the **ilium.** An **ischium** forms the caudoventral portion of the girdle; an unossified **pubis,** the cranioventral portion.

The **femur** (Fig. 1-5) is the bone of the thigh, and the shank contains a tibia and fibula that are fused to form a **tibiofibula.** The two proximal **tarsal bones** in the ankle (**astragalus** or **talus,** and **calcaneus**) are very long and actually add an extra segment to the limb. Several small **distal tarsals** lie distal to them. **Metatarsals** form the sole of the foot, and **phalanges** form the digits distal to the sole. A small spur of bone supports the prehallux.

Pelvic girdle and limb structure are well adapted for giving a powerful, synchronous thrust of both hind limbs

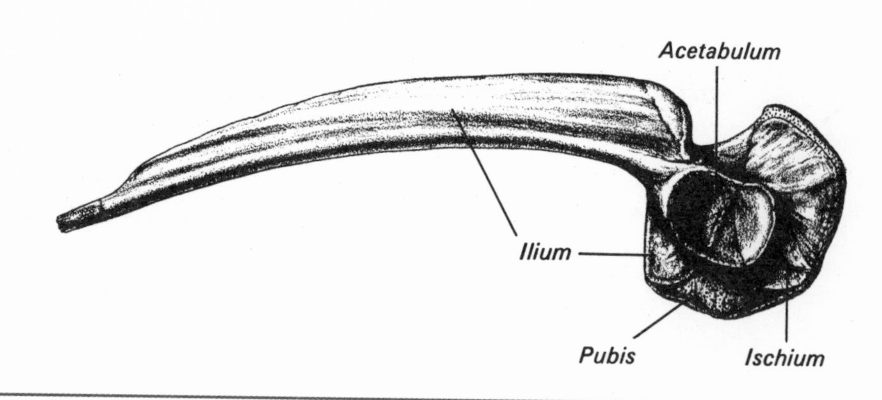

FIGURE 1-10
Lateral view of the pelvic girdle of *Rana*. [After Gaupp.]

FIGURE 1-11
Stages in the take-off and landing of a bullfrog during a strong leap. An arrow indicates the position of the sacroiliac or sacrourostyle joint, and the positions of the ilium and presacral vertebrae are shown by lines. [Modified from stroboscopic photographs made by Mr. Earl Edmiston.]

in swimming and in jumping (Fig. 1-11). When at rest on land, a ranid frog has its hind legs folded up beneath it (or flexed), and its pelvic girdle and urostyle are also flexed, that is, rotated ventrally and forward to some extent at the sacroiliac or sacrourostyle joint. A leap is caused primarily by the rapid straightening out, or extension, of the hind legs, which gives a strong propulsive thrust. In a strong leap, an extension of the pelvic girdle and urostyle gives an additional caudal thrust. These elements can rotate through an arc of about 60°, The continued extension of the girdle may even cause the animal to be "hollow-backed" shortly before landing. A frog lands upon its front legs, which take up much of the shock, but forward inertia is also partly absorbed by the flexion of the pelvis and hind legs at this time. The musculature involved in these movements is described in Exercise 2.

Muscles

MUSCLES ARE THE LARGEST PORTION of a vertebrate's body. By holding the bones in proper relationship to each other, they perform an important function in supporting the body. They are responsible for the movement of the body as a whole, the passage of food down the digestive tract, the ventilation of the lungs, the circulation of the blood, and the movements of most of the materials in the body.

Muscle tissue consists of elongated cells or fibers containing ultramicroscopic protein filaments that can slide past one another, thereby causing the cells to shorten and develop tension. In most muscle fibers the protein filaments, known as myofilaments, are grouped in such a way that each fiber has a cross-banded, or striated, appearance when viewed with a light microscope. **Striated muscle,** as this type of tissue is called, is usually associated with skeletal elements. It is the striated muscles that are considered in this exercise. Usually their action is under voluntary control. Muscle tissue in the walls of blood vessels and most of those of the digestive tract are described as **smooth muscle,** for their cells do not have the cross-banded appearance. A special type of striated muscle, **cardiac muscle,** forms the musculature of the heart wall. Smooth and cardiac muscles are innervated by the autonomic nervous system (Exercise 6) and are involuntary.

When the evolution of muscles is considered, it is convenient to divide the muscular system into **somatic muscles,** which lie in the body wall and appendages, and **visceral muscles** of the gut tube and other visceral organs. Certain of the visceral muscles, the **branchiomeric muscles,** attach onto the visceral arches. These, naturally, are less conspicuous in terrestrial vertebrates than in fishes, but jaw muscles and certain muscles of the throat and shoulder are branchiomeric in origin.

Because it is not possible to consider all of these muscles in an exercise of this scope, emphasis is upon the superficial muscles of the body, but all those of the thigh and shank are described because these regions are usually studied in more detail.

A. MUSCLE DISSECTION AND TERMINOLOGY

Strip the skin from half the body. This is easy to do because the skin is attached to underlying muscles only along certain restricted lines; most of it is separated from the muscles by large **subcutaneous lymph sacs** (Exercise 4). As the skin is removed, notice its rich vascular supply. The skin of the frog is one important site for gas exchange between the blood and the environment. In some regions of the body, it will be necessary also to remove tough sheets of connective tissue, known as **fascia,** which bind certain groups of muscles together. Before beginning to identify the muscles of a particular region (for example, the thigh), separate them by carefully picking away connective tissue from between them. Do not try to cut muscles apart. Take care to observe the direction of the groups of muscle fibers within a muscle as you dissect them. Fibers of a particular muscle, or major part of a muscle, will run in a common direction, and those of an adjacent muscle will have a different direction.

An individual muscle consists of many striated muscle fibers that are held together by their investing connective tissue and have a common attachment. Usually each end of a muscle attaches to a bone or cartilage, but sometimes a muscle attaches onto another muscle or onto a sheet of fascia. Attachments are made by the investing connective tissue, which often is conspicuous enough to form a cordlike **tendon,** or a sheetlike **aponeurosis.** When the muscle fibers come so close to a bone that the investing connective tissue is not grossly apparent, the attachment is said to be **fleshy.**

One end of a muscle, its **origin,** attaches onto a skeletal element that is held in a fixed position when the muscle contracts; the other end, its **insertion,** attaches onto a part that is free to move. It is conventional to consider the proximal end of a limb muscle as its origin and its distal end as its insertion. Origin and insertion are convenient descriptive terms, but their importance should not be exaggerated. The tension developed by a muscle is the same at each end, and in some cases either end of the muscle may be the fixed end, depending on the action being performed.

Muscles perform work only by contracting. A particular muscle will move a limb, for example, in only one direction; the contraction of an antagonistic muscle is required to move the limb back to its original position. Terms that describe the major antagonistic actions are defined here: (1) **Flexion** is a movement that causes a bending, or decrease in the angle, between two parts. The term is most appropriately applied to movements at the elbow, wrist, knee, ankle, and digits. However, bending of the head or trunk toward the ventral side is also called flexion, and lateral bending of the trunk is known as lateral flexion. **Extension** is the antagonistic action. (2) **Protraction** is a forward movement of the entire limb at the shoulder or hip; **retraction** is the opposite movement. (3) **Adduction** is the movement of a part toward a point of reference; the movement away from the reference point is **abduction.** For limb movements, the point of reference is the midventral line of the body; hence movement of the limb ventrally is adduction.

Our knowledge of the actions of frog muscles has been inferred largely from observing their attachments and pulling upon them in cadavers. In living animals, muscles usually interact in complex ways. Flexing your forearm, for example, not only involves contraction and shortening of the biceps brachii muscle, but also the development of tension (but not necessarily shortening) in several trunk and shoulder muscles that hold the scapula and upper arm in place.

Although students often are overwhelmed by the names of muscles, the names can be very helpful, for they generally are descriptive of a muscle's attachments (iliofibularis), shape (deltoideus), number of divisions (triceps), function (adductor magnus), or position (external oblique). Many terms are available for individual muscles. Those selected here are the ones most widely used in the research literature dealing with lower vertebrates. Some of these are terms for human muscles that have been applied to comparable muscles in the frog, but the homology is not always as exact as this practice implies. Important synonyms used in some textbooks are given in parentheses.

The Muscles of the Pelvis and Hindlimb

B. THE MUSCLES OF THE PELVIS AND HINDLIMB

1. The Superficial Muscles of the Thigh

The cranial portion of the thigh (Figs. 2-1, 2-2, 2-3, and 2-4) is covered by a large muscle mass that can be divided into three parts—the **glutaeus magnus (vastus externus),** the **tensor fasciae latae (rectus anticus),** and the **cruralis (vastus internus),** which is partly covered by the first two. The tensor fasciae latae arises from the cranial part of the ilium. The glutaeus magnus arises from more caudal parts of the ilium and, in *Xenopus,* from the sacral fascia. The cruralis arises from the connective tissue capsule of the hip joint. The tensor fasciae latae attaches to the other two, which in turn insert together onto the proximal end of the tibiofibula. For this reason, they are sometimes described as separate heads of a single muscle, the **triceps femoris.** Extension of the shank, an important motion in jumping, is their primary function, but the glutaeus magnus and tensor fasciae latae, which extend across the hip joint, can also help to protract the thigh.

The rest of the dorsal surface of the thigh is occupied by a slender **iliofibularis,** a large **semimembranosus,** and a **gracilis minor** (Figs. 2-1 and 2-3). All of these arise from the caudodorsal portion of the pelvic girdle and insert upon the proximal end of the tibiofibula. By pulling

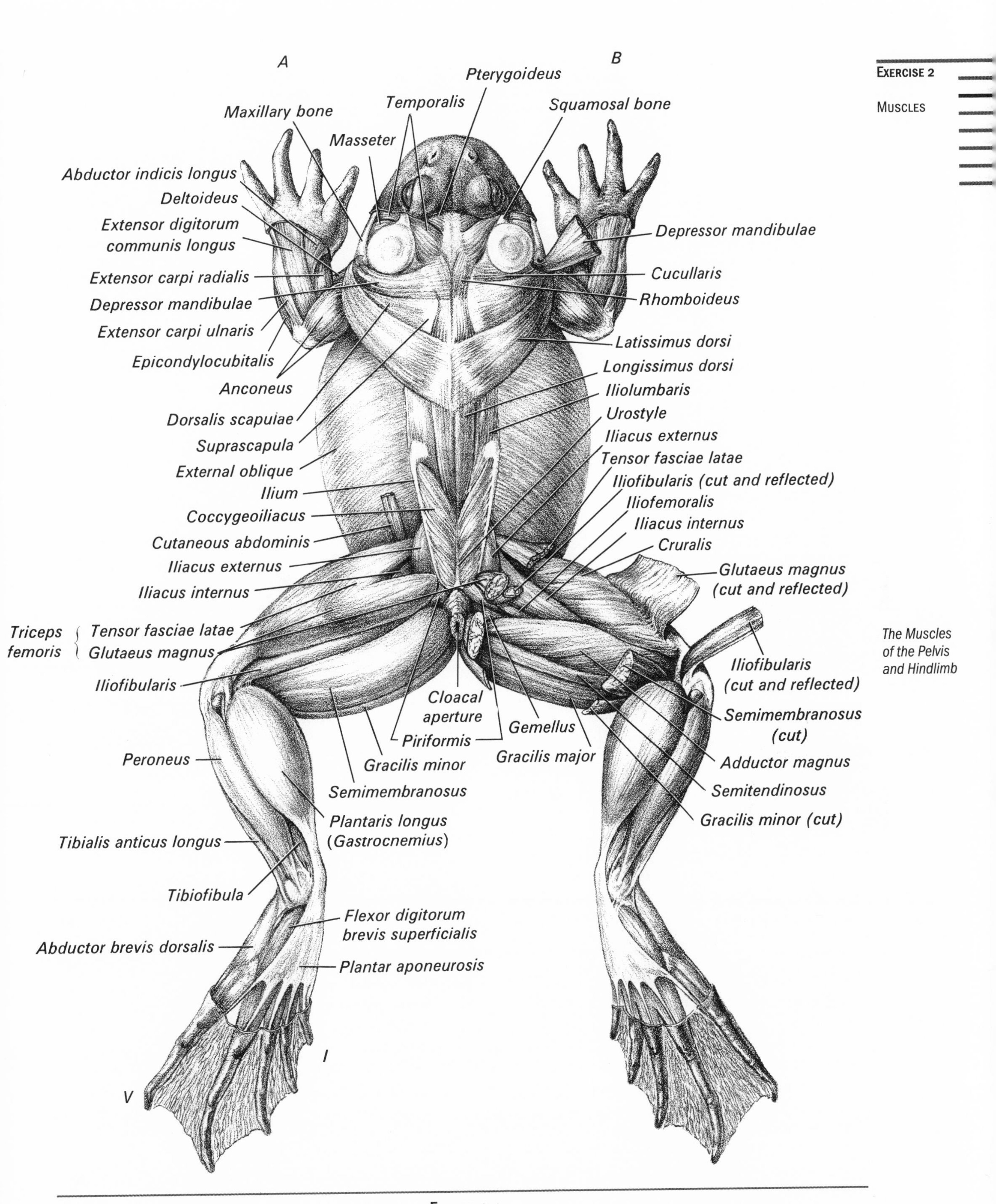

The Muscles of the Pelvis and Hindlimb

FIGURE 2-1
Muscles of *Rana* as seen in a dorsal view. Left side (A), superficial muscles; right side (B), deeper muscles.

The Muscles of the Pelvis and Hindlimb

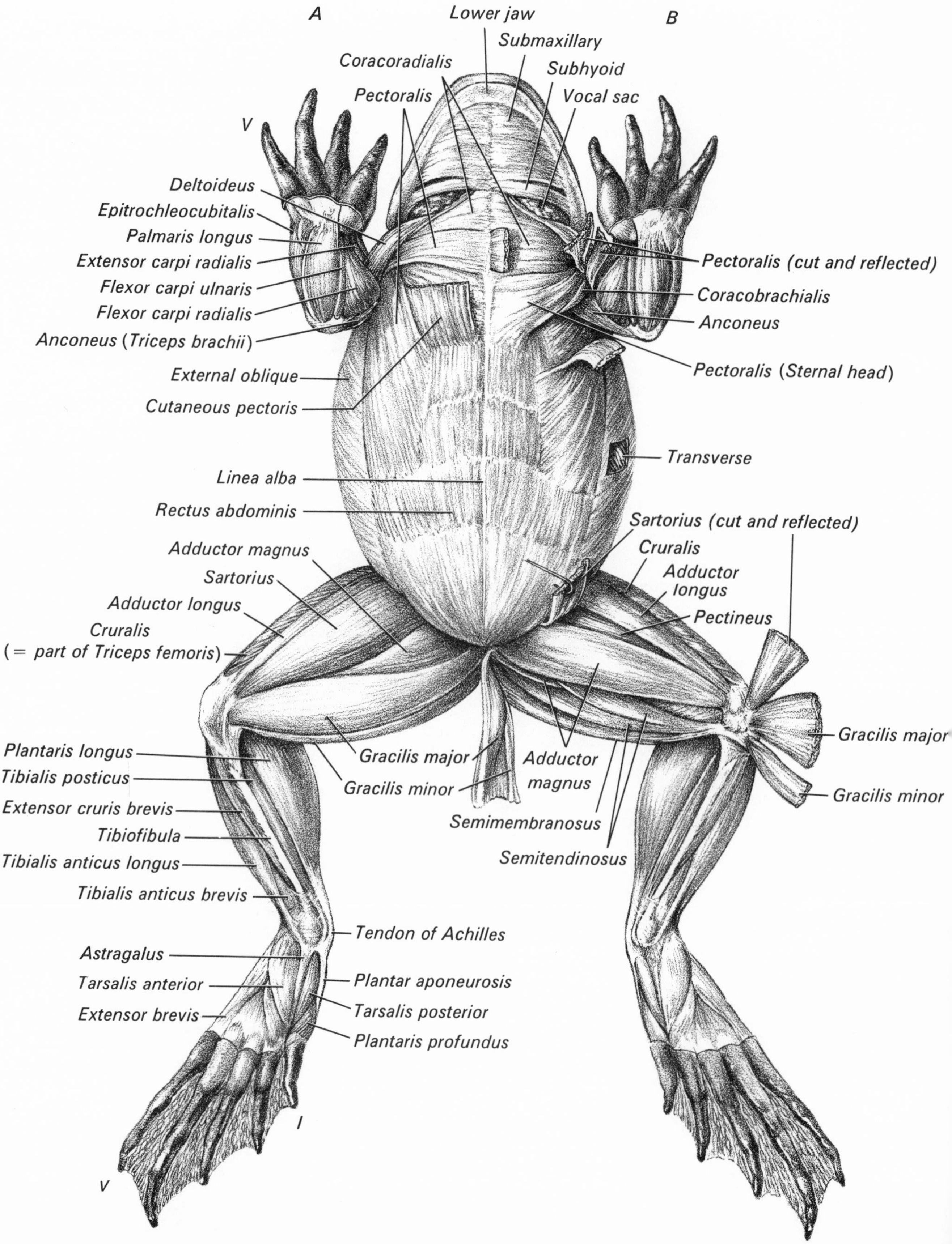

A
Lower jaw
B
Submaxillary
Coracoradialis
Subhyoid
Pectoralis
Vocal sac
V
Deltoideus
Epitrochleocubitalis
Palmaris longus
Extensor carpi radialis
Flexor carpi ulnaris
Flexor carpi radialis
Anconeus (Triceps brachii)
Pectoralis (cut and reflected)
Coracobrachialis
Anconeus
Pectoralis (Sternal head)
External oblique
Cutaneous pectoris
Transverse
Linea alba
Rectus abdominis
Sartorius (cut and reflected)
Cruralis
Adductor magnus
Adductor longus
Sartorius
Pectineus
Adductor longus
Cruralis
(= part of Triceps femoris)
Gracilis major
Plantaris longus
Tibialis posticus
Gracilis major
Adductor magnus
Gracilis minor
Gracilis minor
Extensor cruris brevis
Semimembranosus
Tibiofibula
Semitendinosus
Tibialis anticus longus
Tibialis anticus brevis
Tendon of Achilles
Astragalus
Plantar aponeurosis
Tarsalis anterior
Tarsalis posterior
Extensor brevis
Plantaris profundus
I
V

FIGURE 2-2
Ventral view of the muscles of *Rana*. Left side drawing (A), superficial muscles; right side (B), deeper muscles.

on them you can see that they act primarily to flex the shank, but during shank extension their contraction probably helps to retract and abduct the thigh.

Study the ventral surface of the thigh (Figs. 2-2 and 2-4) and find the cruralis again in this view. The broad, superficial muscle on the ventral surface of the thigh caudal to the cruralis is the **sartorius.** In *Rana* the sartorius arises from the cranioventral part of the pelvic girdle; in *Xenopus,* it originates from the caudal border of the abdominal muscles. In both genera, it extends diagonally, laterally, and caudally to insert on the proximal end of the tibiofibula. It helps to flex the shank and to adduct and protract the thigh. An adductor complex of muscles lies deep to the sartorius, and parts of it can be seen in *Rana* lying just cranial and caudal to the sartorius. The adductors are described below.

Again find the gracilis minor on the caudal border of the thigh. A large **gracilis major** lies just cranial to it on the ventral surface of the thigh. It arises from the caudoventral portion of the pelvic girdle, and inserts with the gracilis minor on the proximal end of the tibiofibula. It helps to flex the shank and to retract and adduct the thigh. In *Xenopus* a semitendinosus lies superficially between the the sartorius and gracilis major, but this muscle is more deeply situated in *Rana.* It will be considered later.

2. The Muscles of the Pelvis and the Deep Muscles of the Thigh

The Muscles of the Pelvis and Hindlimb

A number of short muscles extend from the pelvic girdle to the proximal end of the femur. Although several of these are visible in a superficial view, many of the superficial muscles on the surface of the thigh must be cut through and turned back, or reflected, to see these short muscles clearly. First, cut through the tensor fasciae latae, glutaeus magnus, iliofibularis, semimembranosus, and gracilis minor near their insertions, but leave a short section of each muscle at its insertion so that the relationships of the muscles to each other remains apparent (Figs. 2-1B and 2-3B).

An **iliacus externus** extends from the iliac blade to the proximal end of the femur, and a deeper **iliacus internus** extends from the ilium distally to the central portion of the femur. The iliacus externus is a very wide muscle in *Xenopus* and covers the iliacus internus. Both of these muscles are important protractors of the thigh, and they are assisted by a small **iliofemoralis,** which lies on the caudal side of the iliacus internus and extends from the ilium to the femur.

Two other small muscles, the **piriformis** and **gemellus,** lie medial and caudal to the iliofemoralis. The piriformis, which is reduced in size and sometimes absent in *Xenopus,* arises from the urostyle, and the gemellus, from the caudodorsal part of the ischium, and they insert upon the proximal part of the femur. Their caudal position makes them retractors of the thigh. A very short **obturator internus** lies beneath the gemellus. It extends from the pelvis to the femur across the caudodorsal surface of the hip joint, which it helps to support.

Turn your specimen over, and cut through and reflect the sartorius and the gracilis major (Figs. 2-2B and 2-4B). The adductor complex of muscles can now be seen. In *Rana,* a thin **adductor longus** lies between the sartorius and cruralis. It arises from the cranioventral part of the pelvic girdle and inserts on the distal end of the femur. *Xenopus* does not have this muscle. In *Rana* a larger **adductor magnus** lies deep to and caudal to the sartorius, but in *Xenopus* it is entirely covered by the sartorius. It arises from the caudal part of the pelvic girdle by distinct ventral and dorsal heads. The dorsal head can be seen most clearly from the dorsal side of the thigh (Figs. 2-1B and 2-3B), where it lies deep to the origin of the semimembranosus and caudal to the piriformis and gemellus. The two heads of the adductor magnus converge and insert in common on the distal one third of the femur.

A large **semitendinosus** lies caudal to the adductor magnus (Fig. 2-2B and 2-4B). It is covered by the gracilis major in *Rana,* but is more superficial in *Xenopus.* It arises by dorsal and ventral heads from the ventral and caudal parts of the pelvic girdle. The ventral head passes between the two heads of the adductor magnus in *Rana.* The semitendinosus inserts by means of a conspicuous tendon on the proximal end of the tibiofibula. It acts both to flex the shank and retract the thigh.

A small, triangular-shaped **pectineus** lies on the ventral surface of the thigh between the cruralis and the ventral head of the adductor magnus. It arises from the ventral part of the pelvic girdle, inserts upon the proximal half of the femur, and assists in adducting the thigh. Push the ventral head of the adductor magnus caudally (or cut through it; Fig. 2-5). In *Rana* two additional small muscles, the **obturator externus** and the **quadratus femoris,** arise from the pelvis caudal to the pectineus and insert upon the proximal portion of the femur. They too are primarily adductors. The gemellus, described above, is dorsal to the quadratus femoris. These two small muscles are united in *Xenopus.*

3. The Muscles of the Shank

The large muscle forming the calf of the shank, or lower leg, (Fig. 2-1, 2-2, 2-3, and 2-4) is the **plantaris longus** (often called **gastrocnemius,** although it is only partly comparable to this human muscle). It arises by means of large dorsal and ventral tendons from the distal end of the femur, and terminates in the powerful **tendon of Achilles,** which passes around the ankle and spreads out on the sole of the foot to form the **plantar aponeurosis.** Several small foot muscles arise from the deep surface of this aponeurosis, which then forms five tendons, one going to the terminal segment of each toe. The plantaris longus, which is the most powerful extensor of the foot (extension being an

The Muscles of the Pelvis and Hindlimb

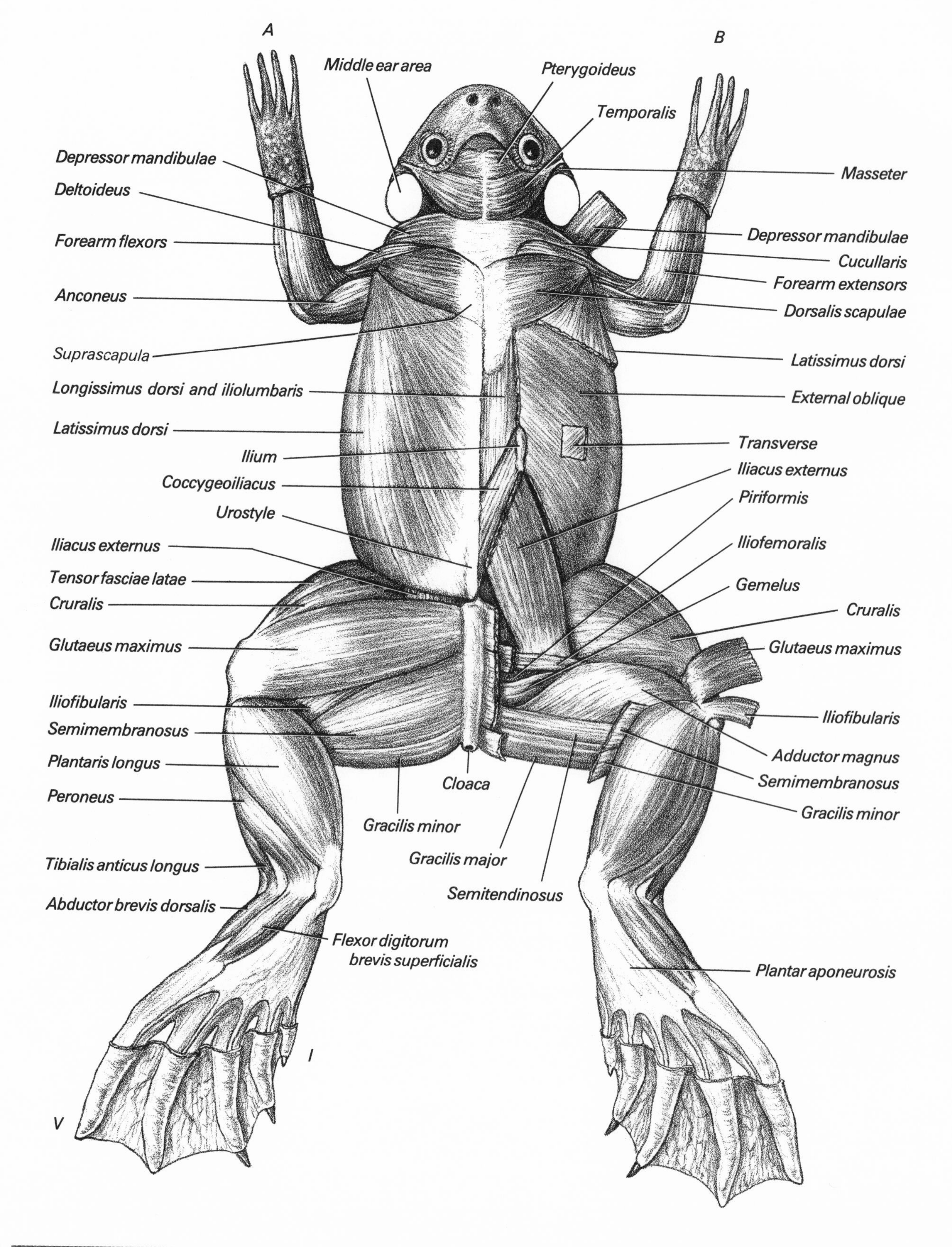

FIGURE 2-3
Muscles of *Xenopus* as seen in a dorsal view. Left side (A), superficial muscles; right side (B), deeper muscles.

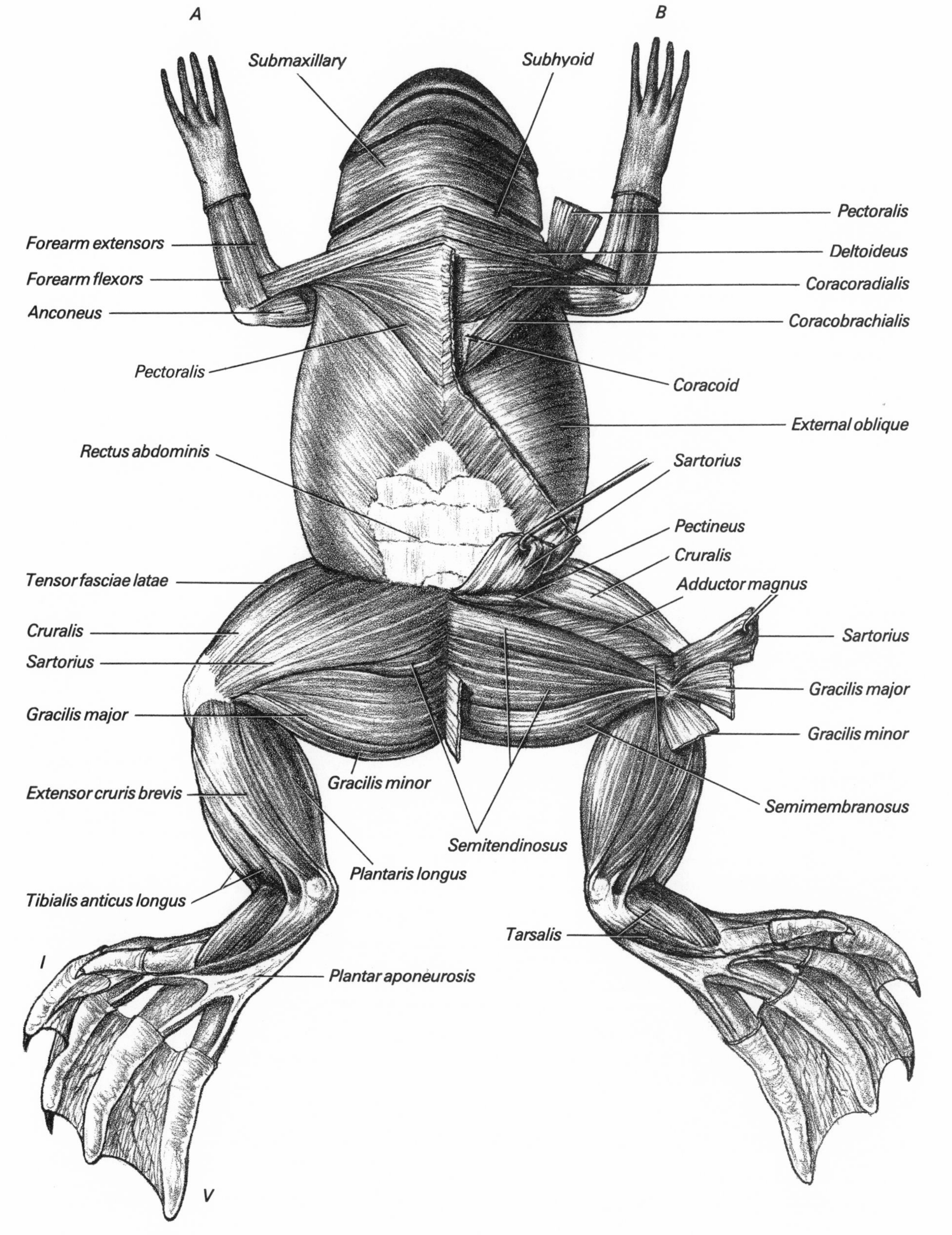

The Muscles of the Pelvis and Hindlimb

FIGURE 2-4
Muscles of *Xenopus* as seen in a ventral view. Left side (A), superficial muscles; right side (B), deeper muscles.

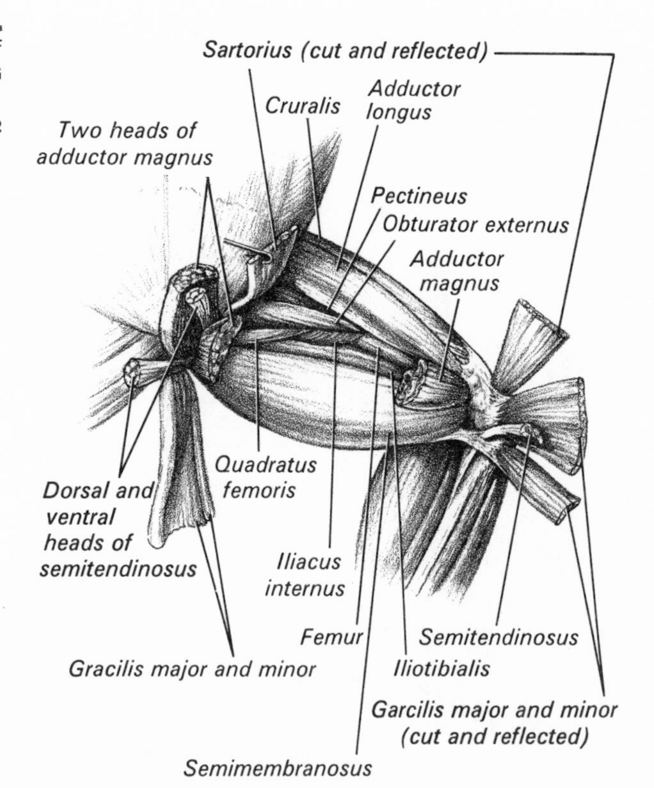

Figure 2-5
Ventral view of the deepest thigh muscles of *Rana* after reflection of the adductor magnus and semitendinosus (see Fig. 2-2B).

action often called plantar flexion), plays a very important role in jumping. It also assists in flexing the toes. A **tibialis posticus,** best found from the ventral side, lies beneath the plantaris longus. It is visible in a ventral view of *Rana,* but is completely covered by the plantaris longus in *Xenopus.* It arises from most of the tibiofibula shaft, inserts by a short tendon onto the proximal end of the astragalus, and assists in foot extension.

Three conspicuous muscles occupy the craniolateral surface of the shank. The **peroneus** is the largest and can be seen best from the dorsal side, where it is located lateral to the plantaris longus (Figs. 2-1 and 2-3). The **tibialis anticus longus** lies lateral and ventral to the peroneus and can be seen most clearly from the ventral surface (Figs. 2-2 and 2-4). The **extensor cruris brevis** lies medial to the tibialis anticus longus on the ventral surface. These three muscles have rather long tendons that cross the front of the knee joint, beneath the tendon of the triceps femoris, to arise from the distal end of the femur. The extensor cruris brevis inserts on the tibiofibula and is an extensor of the shank. It is particularly well developed and long in *Xenopus.* The peroneus inserts by a tendon on the proximal tarsal bones. The tibialis anticus longus divides into two bellies distally that insert by conspicuous tendons on the dorsal and ventral surfaces of the tarsus. Peroneus and tibialis anticus longus are primarily flexors (dorsiflexors) of the foot, but since they also extend across the knee, they can help extend the shank. A fourth and smaller muscle, the **tibialis anticus brevis,** arises from the distal portion of the tibiofibula, inserts upon the tarsus, and is a flexor of the foot.

C. The Muscles of the Shoulder and Forelimb

The Muscles of the Shoulder and Forelimb

Before the major muscles of the shoulder and arm can be identified it is necessary to find and reflect the **depressor mandibulae,** a strap-shaped jaw muscle located just caudal to the tympanic membrane on the top of the head (Figs. 2-1 and 2-3). Although *Xenopus* has no tympanic membrane, it has a roundish middle ear region in a comparable position. The depressor mandibulae arises from fascia on the top of the head, inserts on the caudal end of the lower jaw, and opens the jaw. Detach its origin and reflect it. The **cucullaris** is the most cranial of the dorsal shoulder muscles. It lies beneath the depressor mandibulae and extends from the back of the skull to the proximal end of the humerus. The cucullaris is a branchiomeric muscle that has secondarily acquired an attachment onto the humerus. A **dorsalis scapulae** lies caudal to the cucullaris and arises from the dorsolateral surface of the scapula. A **latissimus dorsi** arises from the fascia of the back caudal to the dorsalis scapulae. Both of these muscles insert on the proximal end of the humerus. The latissimus dorsi is exceptionally wide in *Xenopus* and covers most of the back and side of the abdomen. Cucullaris, dorsalis scapulae, and latissimus dorsi all abduct the humerus. The more cranial of these muscles also protract it and the more caudal ones retract it.

Examine the ventral muscles of the shoulder region (Figs. 2-2 and 2-4). *Rana* has a superficial **cutaneous pectoris** that extends from the ventral abdominal wall forward to insert on the skin. Free and reflect it, for it covers some of the chest muscles. *Xenopus* lacks this muscle. A large fan-shaped **pectoralis,** which can be divided into several parts, covers most of the chest. It arises from the abdominal muscles and sternum and inserts on the proximal end of the humerus. A small **coracobrachialis** lies deep to the caudal border of the pectoralis. It arises deep to the pectoralis from the coracoid bone and inserts along the humerus. A **deltoideus** covers the front of the shoulder, where it arises from the cranial end of the sternum and pectoral girdle and inserts along the humerus distal to the insertion of the pectoralis. It is longer in *Xenopus* than in *Rana.* Pectoralis, coracobrachialis, and deltoideus all adduct the arm. The cranial parts of this musculature also protract the arm and the caudal ones retract it.

A **coracoradialis** can be seen on the ventral surface between the pectoralis and the deltoideus. It arises from the coracoid deep to the pectoralis and forms a powerful tendon that extends along the ventral surface of the humerus to insert on the radioulna. It is the major flexor of the forearm, but it also adducts the entire arm, for it crosses the shoulder as well as the elbow joint.

The **anconeus (triceps brachii)** covers the dorsal surface of the humerus. It arises by two heads from the humerus and one from the scapula and forms a powerful tendon that cross the elbow joint to insert on the radioulna. It is primarily an extensor of the forearm.

The muscles in the forearm and in the hand can be sorted into a group of **extensors** on the dorsolateral surface of the forearm, which extend the hand and the digits, and a group of antagonistic **flexors** on the ventromedial surface. On the lateral edge of the forearm these two groups are separated by the insertion of the anconeus; on the medial edge, the deltoid and tendon of the coracoradialis pass between them. The individual extensors and flexors are difficult to see in an animal as small as *Xenopus,* but they can be recognized in large specimens of *Rana.* The superficial extensors are, from the medial edge to the lateral edge, the **extensor carpi radialis,** the **abductor indicis longus,** the **extensor digitorum communis longus,** the **extensor carpi ulnaris,** and the **epicondylocubitalis** (Fig. 2-1). The superficial flexors are, from medial to lateral, the **flexor carpi radialis,** the **flexor carpi ulnaris,** the **palmaris longus,** and the **epitrochleocubitalis** (Fig. 2-2).

D. THE MUSCLES OF THE TRUNK

The **cutaneous pectoris,** which attaches into the skin, has been seen overlying the pectoralis (Fig. 2-2). Another cutaneous muscle, the **cutaneous abdominis,** extends from the caudal part of the abdominal musculature to the skin on the back (Fig. 2-1). It too is absent in *Xenopus.* The contraction of these muscles probably compresses certain subcutaneous lymph sacs and may possibly help to circulate the lymph, but little else is known about their function.

A longitudinal **rectus abdominis** (Figs. 2-2 and 2-4), located on each side of the midventral line, forms the ventral portion of the abdominal wall. Traces of the segmentation of the muscles, which is characteristic of the trunk muscles of fishes, can still be seen in it. The rest of the abdominal wall is formed by two thin sheets of muscles, the **external oblique** and **transverse** (Figs. 2-2 and 2-3), whose fibers run nearly perpendicular to each other. The transverse can be seen if a small window is cut through the external oblique.

Muscles on the back of *Rana* can be seen by removing the fascia on either side of the middorsal line (Fig. 2-1). In *Xenopus,* you must reflect the latissimus dorsi, the dorsal part of the external oblique, and the iliacus externus because these muscles cover those of the back (Fig. 2-3). A **cocygeoiliacus** extends diagonally from the urostyle cranially and laterally to attach to the iliac blade. Cranial to this, the back muscles of *Rana* form two longitudinal columns: a median **longissimus dorsi** and a **lateral iliolumbaris.** These two muscles are united in *Xenopus.* Besides bracing the back, these three muscles appear to extend the urostyle and pelvic girdle, an action that takes place during a strong leap. Contraction of the ventral abdominal musculature restores the urostyle and girdle to its flexed, resting position. (Exercise 1).

Several small trunk muscles extend from the body axis to the scapula and suprascapula, which they help to hold in place. One of these, the **rhomboideus,** is visible in a superficial dissection of the back (Fig. 2-1B).

The Muscles of the Head

E. THE MUSCLES OF THE HEAD

Muscles on the dorsal surface of the head act on the jaws (Figs. 2-1 and 2-3). The **depressor mandibulae** was seen and reflected when you dissected the shoulder muscles. A **temporalis** arises on the top of the head and passes ventrally just cranial to the tympanic region. A small **masseter** lies caudal to the distal part of the temporalis, and a **pterygoideus** arises from the skull just in front of the origin of the temporalis. All insert on the lower jaw and pull it closed.

The **submaxillary (mylohyoid)** and **subhyoid** (Figs. 2-2 and 2-4), which extend transversely on the ventral surface of the head, are important in swallowing and in the olfactory and breathing movements of the floor of the mouth and the floor of the pharynx (Exercise 3). All of the head muscles are branchiomeric in origin.

Digestive and Respiratory Systems

All animals must take into their bodies a variety of materials to sustain their metabolism. These include food, vitamins, mineral ions, water, and oxygen. Most of the water enters through the thin vascular skin of frogs; the skin is also important in the uptake of ions and in gas exchange. Most of the oxygen enters, and much of the carbon dioxide leaves, through the lungs; other materials enter through the digestive system. Roughage and other undigested material, bacteria, and bile pigments derived from the breakdown of hemoglobin in the liver are discharged as feces by the digestive system. Nitrogenous wastes of metabolism are removed by the excretory system (Exercise 5).

In all vertebrates, the respiratory system develops embryonically as outgrowths of the pharyngeal part of the digestive tract. It is therefore convenient to study these two systems at the same time.

A. THE BUCCOPHARYNGEAL CAVITY

Open the **mouth** by cutting directly back through the angle of the jaw on each side. Extend the cut caudally nearly to the shoulder. Pull the floor of the mouth ventrally, thereby exposing the buccal cavity and the pharynx (Figs. 3-1 and 3-2). The **buccal,** or **oral, cavity** is the cavity of the mouth and is bounded by the jaws; the **pharynx** is a transitional region between the buccal cavity and the gullet, or **esophagus,** which leads to the stomach. The pharynx is more clearly demarcated in an embryo because the embryo has gill pouches in this region. In an adult, the entrance of the **auditory,** or **eustachian, tubes,** which have developed from the first pair of gill pouches, indicate the approximate cranial border of the pharynx. In *Rana,* an auditory tube lies on either side of the pharynx roof, but the tubes of *Xenopus* have a common opening in the midline of the pharynx roof. Auditory tubes lead to the middle ear cavities (Exercise 7).

A large **tongue** lies on the floor of the buccal cavity in *Rana.* It is attached to the floor of the mouth near the chin in such a way that its forked caudal end can be quickly flicked out of the mouth toward an insect or other small animals upon which the frog feeds (Fig. 3-3). As it is thrust out, the tongue scrapes off a sticky secretion produced by glands in the roof of the mouth, so the prey sticks to the tongue and is pulled back into the mouth. Like larval amphibians and fishes, *Xenopus* lacks a muscular tongue. *Xenopus* uses its arms to help push food into lts small mouth.

Rana has two patches of **vomerine teeth** in the roof of the oral cavity. These are absent in *Xenopus.* In both species, the lower jaw has no teeth and those of the upper jaw are very small. Teeth are used simply to hold the prey in the mouth while it is being swallowed whole. Amphibians do not masticate their food.

Paired nasal passages open into the roof of the buccal cavity by **internal nostrils,** or **choanae.** In *Rana,* one is lateral to each patch of vomerine teeth. Air enters the lungs by way of the **glottis,** a short longitudinal slit in the floor of the pharynx.

In a male *Rana,* look for the openings of the paired **vocal sacs,** one on each side of the floor of the pharynx

The Buccopharyngeal Cavity

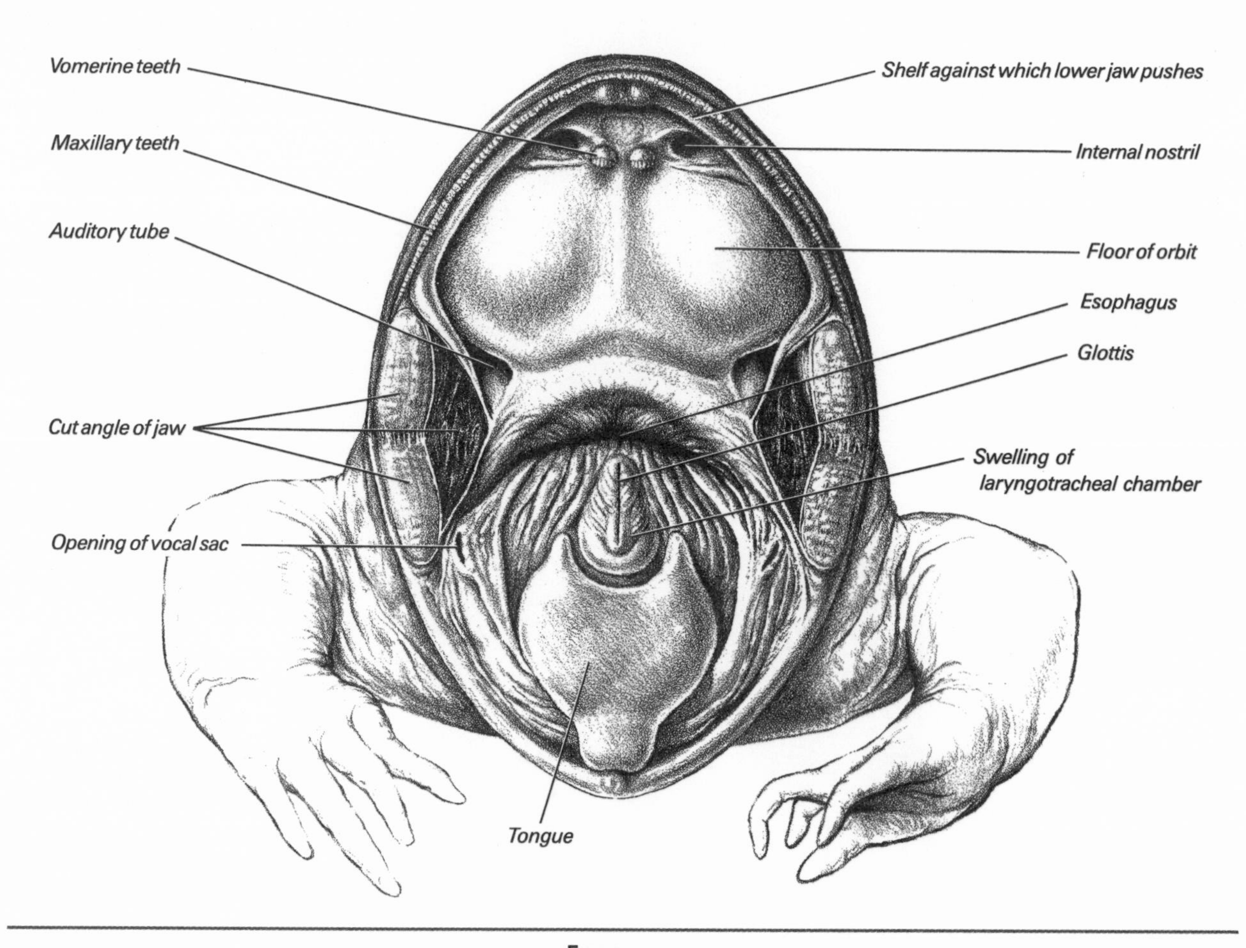

FIGURE 3-1
Interior view of the buccopharyngeal cavity of a male *Rana.*
The angles of the jaw have been cut so that the mouth can be opened wide.

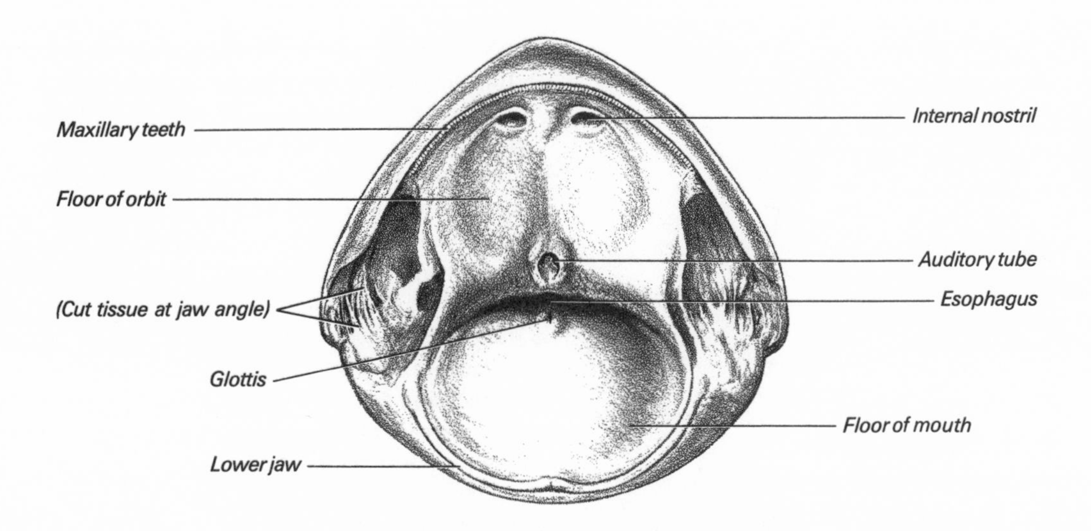

FIGURE 3-2
Interior view of the buccopharyngeal cavity of *Xenopus*.

slightly medial to the angle of the jaw. Most adult male frogs vocalize during the breeding season, and females of the same species are attracted by their distinctive calls. The vocal sacs serve as resonating chambers. *Xenopus* lacks vocal sacs.

The buccopharyngeal cavity is kept free of debris by the continual movement of a sheet of mucus back into the esophagus. The mucus is secreted by certain cells in the lining of the cavity and is carried by other ciliated cells. This mechanism can be demonstrated on a pithed frog by placing small particles on the roof of the buccopharyngeal cavity.

B. THE BODY CAVITY

If the trunk has not been already skinned for the study of the muscles, make a longitudinal incision that is slightly to the left of the midventral line and extends from the pectoral to the pelvic girdle (Fig. 3-4). Now cut through the muscle layers of the abdominal wall in the same manner. Lift up the wall, find the large **ventral abdominal vein** on its inner surface, and carefully pull it away from the midventral line. Small tributaries draining the body wall will have to be cut. Cut across the middle of the ventral abdominal vein so that it can be pushed aside. Next, make transverse incisions through the skin and muscle just cranial to each hind leg, and caudal to each front leg. Continue the cuts to the back, and turn back the flaps of the body wall. The flaps may be pinned down in a dissecting pan. Finally, lift up the pectoral girdle, carefully free the structures underneath that adhere to it, and make a longitudinal cut forward through it.

The Body Cavity

The body cavity, or **coelom,** in which the visceral organs are located, is now exposed. It is completely lined by a shiny layer of coelomic epithelium. The epithelium lining the abdominal wall is called the **parietal peritoneum;** that covering the walls of the viscera, the **visceral peritoneum.** Bits of the coelomic epithelium may be peeled from the abdominal and visceral walls. Membranes of coelomic epithelium, the **mesenteries,** extend both from the middorsal line and from part of the midventral line of the body wall to the organs, and they also extend between many of the organs. These relationships are shown in Figure 3-5. The coelom is a space that facilitates the expansion, contraction, and other functional movements of the viscera. Mesenteries limit these movements to some extent, and are pathways through which blood vessels and nerves travel.

The coelom of the frog is divided into two chambers: a large **pleuroperitoneal cavity** containing the lungs and most of the other viscera, and a small **pericardial cavity** enclosing the heart. The pericardial cavity lies just dorsal to the pectoral girdle, and it was probably cut open at

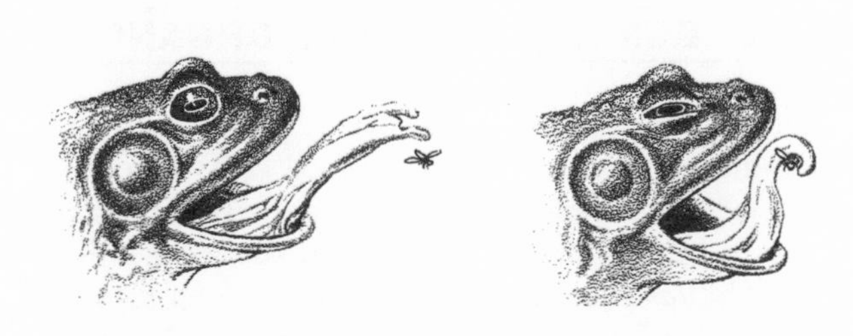

FIGURE 3-3
Tongue action of *Rana*. [After Gadow].

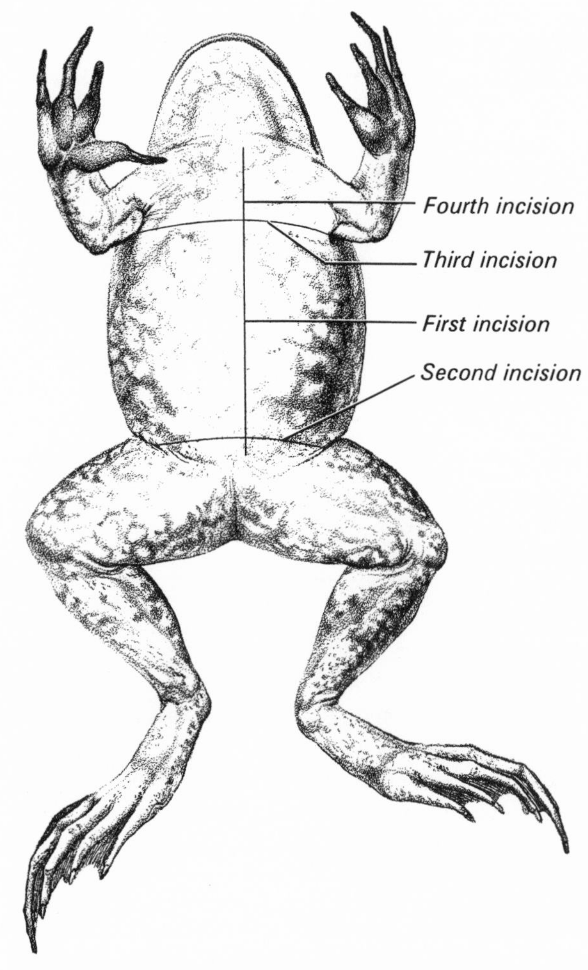

FIGURE 3-4
Ventral view showing the sequence of incisions that should be made to open the body cavity.

the time the girdle was cut. The coelomic epithelium that lines the pericardial cavity is called the **parietal pericardium,** and the heart is covered with **visceral pericardium.**

C. GENERAL VISCERAL ORGANS

Before continuing with a study of the digestive tract, identify the organs in the pleuroperitoneal cavity (Fig. 3-6). The large, dense, lobed organ occupying much of the cranioventral part of the cavity is the **liver.** The liver has three lobes in *Rana* but only two in *Xenopus.* Lift up the right side of the liver and find the **gall bladder,** a small, dark sac attached to the dorsal surface of the liver. The **stomach** lies dorsal to the liver on the left side of the body. One **lung** lies dorsolateral to the stomach; the other, in a comparable position on the other side of the body. The lungs are exceptionally large in *Xenopus* and extend nearly to the hind legs. In *Xenopus* they are attached by mesenteric folds to the lateral body wall and to the lateral edge of the liver. You should cut through these attachments to approach deeper structures. Lift up the caudal part of the stomach. The small, dense, round body on the left side of the mesentery supporting the intestine is the **spleen,** an organ in which blood cells are manutured and stored (Exercise 4). A **small intestine** leads from the caudal end of the stomach. The lobate whitish tissue lying in the loop between the first part of the small intestine and the stomach is the **pancreas.** Follow the coils of the small intestine caudally and they will lead to the wider, **large intestine.** The fingerlike lobes of fat that must be pushed aside are the **fat bodies.** One mass is attached to the cranial end of each testis or ovary. The **testes** of a male are a pair of small, oval bodies, one on each side of the mesentery that supports the intestine. **Ovaries** occupy the same position, but vary greatly in size according to the reproductive state of the female. Shortly before ovulation, they are filled with mature eggs and occupy all available space in the pleuroperitoneal cavity. It may be necessary to remove one ovary in order to see the other organs clearly. An elongated dark **kidney** lies against the back, dorsal to each **gonad** (testis or ovary). In females, a long coiled white tube, the **oviduct,** will be found lateral to each kidney. Males may have a smaller vestigeal oviduct. The **urinary bladder** is a sac located just cranial to the hind legs and ventral to the large intestine. When distended, it is a large, bilobed organ; when contracted, it is much smaller.

The Digestive Tract

D. THE DIGESTIVE TRACT

Pass a probe or a pair of blunt forceps through the pharynx and down the **esophagus.** By feeling with your fingers, you can tell when the probe enters the stomach. Since a frog lacks a neck, the esophagus is quite short. There is no line of demarcation between it and the stomach on the outer surface of these organs, but microscopically there is a change in the type of cell lining their cavities.

The **stomach** is a large, saccular organ in which food is stored and digestion initiated. It curves toward the right side of the body and usually has a J-shape (Fig. 3-6). Cut it open. If it is not completely filled with the remains of food organisms, its lining forms conspicuous longitudinal folds. Its caudal termination is a thick, muscular **pyloric sphincter,** whose contraction keeps food in the stomach until it is partly digested by the gastric juice and broken up mechanically by a churning action. Glands in the stomach wall secrete the enzyme precursor **pepsinogen** and **hydrochloric acid.** In the acid environment, pepsinogen is converted to active **pepsin,** which then initiates the splitting of proteins. Mucus produced by other glands protects the stomach lining from autodigestion.

Digestion is completed and food is absorbed in the small intestine. Its first part, the **duodenum,** curves cranially toward the liver; the rest of it is comparable to the mammalian **jejunoileum.** Slit open a part of the small intestine and observe the irregular folds that increase the internal surface area available for the absorption of digested food.

Most of the secretions that enter the intestinal lumen come from the pancreas and liver, although some proteolytic enzymes and considerable mucus are secreted by individual cells lining the small intestine. The **pancreas** produces enzymes that act on all categories of food (**carbohydrases, lipase, proteinases,** and **nucleases**). The liver produces no enzymes, but the **bile salts** in the bile, which originate in the liver, emulsify fats, thereby facilitating the action of lipase, and also aid in the absorption of glycerol, one of the products of fat digestion. The secretions of the pancreas and liver are also alkaline. By neutralizing the acidity of the gastric contents they produce an environment in which the pancreatic and intestinal enzymes can act.

Bile leaves the liver through inconspicuous **hepatic ducts** that coalesce with a **cystic duct** from the gallbladder to form a **common bile duct.** As the common bile duct passes caudally through the pancreas, it receives several minute **pancreatic ducts.** It finally emerges from the pancreas to enter the duodenum. The pancreas releases its secretions only when food is in the duodenum, but bile, which is secreted continuously, backs up into the gallbladder where some water is reabsorbed and bile is stored until food enters the intestine.

Besides their digestive role, the liver and pancreas have other functions. The liver helps to convert many absorbed foods carried to it by the circulatory system. Excess sugars, fats, and amino acids are stored, largely in the form of glycogen, and the excretory product urea is synthesized from ammonia derived from the deamination of amino acids. Urea is eliminated by the kidneys. The liver is also the site where most of the proteins in the blood plasma are synthesized. The **bile pigments** are excretory products derived from the breakdown in the liver of hemoglobin from red blood cells that are destroyed in the spleen and liver. The bile pigments are responsible for the color of the feces.

Although most of the pancreas produces digestive enzymes, it also contains microscopic clumps of endocrine tissue, **the islets of Langerhans,** which produce the

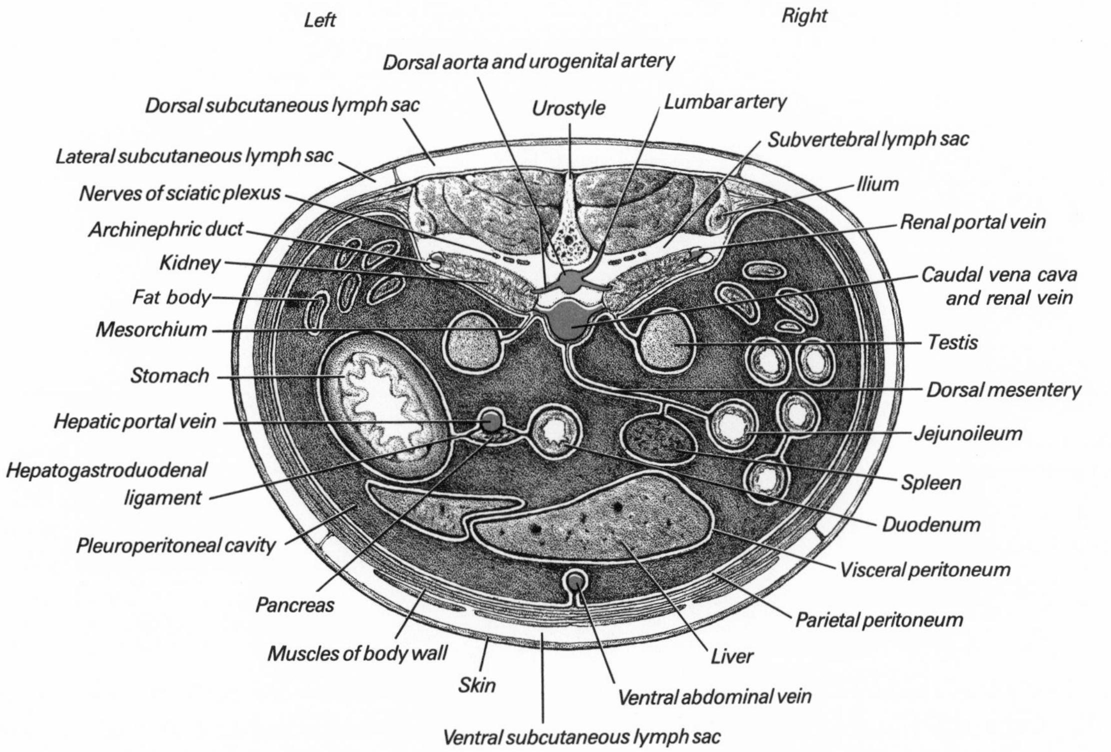

FIGURE 3-5
Diagrammatic cross section through a male *Rana* at the level of the testis. Coelomic epithelium lines the body wall, and double layers of coelomic epithelium extend as mesenteries to the visceral organs, which are also covered by it.

The Digestive Tract

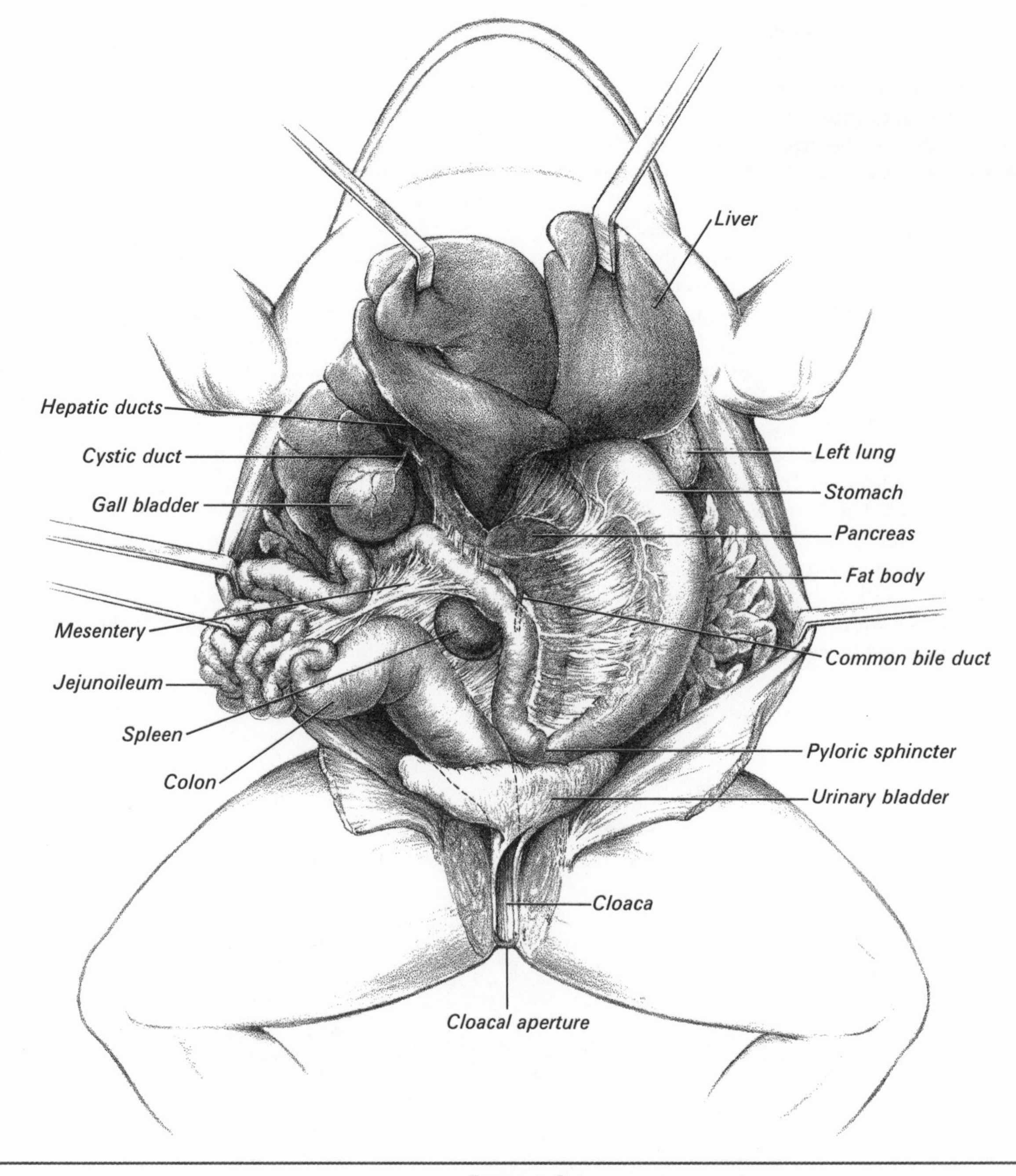

FIGURE 3-6
Ventral view of the digestive tract and associated organs of *Rana*.
The liver has been pulled forward, and the pelvic canal has been cut open.

hormones **insulin** and **glucagon.** Insulin enters the circulatory system and is carried to all parts of the body. It is necessary for the conversion of blood glucose into glycogen in liver and muscle cells and for the metabolism of glucose by the cells of the body. Insulin, therefore, lowers blood sugar levels. Glucagon has an antagonistic effect.

The large intestine, or **colon,** is a short segment from which considerable water and ions are absorbed. The water and ions come not only from the food but also from the digestive secretions that enter the stomach and intestine. Bacteria residing in the colon may break down cellulose into products that can be absorbed. As products are absorbed, the undigested residue and many of the bacteria form feces that are stored temporarily and finally evacuated. The colon has a large diameter cranially, but narrows caudally as it passes through the pelvic canal to enter the **cloaca.** To see this region, carefully cut through the pelvic girdle and spread the hind legs apart. Cut only in the midventral line, and do not injure the urinary bladder. The cloaca is the terminal part of the digestive tract, which also receives the urinary and genital ducts. Notice that the urinary bladder enters it ventrally. You may see other urogenital ducts that enter the cloaca dorsally. The cloaca opens on the body surface at the **cloacal aperture.**

E. THE RESPIRATORY SYSTEM

Pass a probe or a pair of forceps into the **glottis,** and you will be able to feel that the probe soon enters a lung. The glottis is the entrance to a short chamber that is comparable to both our larynx and our trachea and hence is called the **laryngotracheal chamber.** Because frogs have no neck, they have no long windpipe or trachea. A part of this chamber can be seen by dissecting deeply just cranial to the fork formed by a pair of large vessels that emerge from the cranial end of the heart (Fig. 4-3). Be careful not to injure these and other vessels. Enlarge the glottis by making a short longitudinal incision that extends cranially and caudally from it. Spread it open. The longitudinal folds that you see within the laryngotracheal chamber of *Rana,* one on each side, are the **vocal cords** (Fig. 3-7). *Xenopus* does not have them. Beneath the vocal cords you may be able to see the entrance to each lung. The glottis is supported on each side by a small cartilage, which has evolved from one of the gill arches of a fish (Exercise 1). It can be felt with a pair of forceps.

In most preserved specimens, the **lungs** are somewhat contracted. Their true size can be best appreciated by inserting a small glass tube into the glottis of a freshly killed specimen and blowing them up. Remove part of a lung from your specimen and slit it open. It is a sac whose internal surface area has been greatly increased by pocket-shaped folds, which in turn are subdivided by secondary folds.

Gas exchange between the environment and blood occurs through the skin **(cutaneous respiration),** the lining of the mouth and pharynx **(buccopharyngeal respiration),** and the lung **(pulmonary respiration).** The amount of exchange through these different surfaces varies among species, and also within one species; in the frog, the amount of exchange varies according to whether it is on land or submerged in water. Data on the European *Rana temporaria,* which has habits somewhat similar to our *Rana pipiens,* indicate that about 37% of the respiratory capillaries are in the skin, 1% in the buccopharyngeal lining, and 62% in the lungs. Clearly, the largest respiratory surface is in the lungs, and it should be the primary gas exchange surface, except when the animal is completely submerged. However, as important as the size of the lung's gas exchange surface area is the rate of lung ventilation. Because air is 20% oxygen, lung ventilation rate is adequate to supply most of the animal's oxygen needs. Even the highly aquatic *Xenopus* surfaces to breathe air. But the elimination of carbon dioxide presents special problems. Some carbon dioxide, of course, diffuses from the blood into the lungs and is eliminated. However, the rate of lung ventilation is not sufficient to allow enough carbon dioxide to diffuse out of the blood because the carbon dioxide level in the lungs soon equals that in the blood. Most carbon dioxide, therefore, is eliminated through the skin, from which it can be immediately dissipated into the surrounding air or water, where carbon dioxide levels are very low.

Movement of air into and out of the buccopharyngeal cavity and lungs has been studied carefully by de Jongh and Gans (1969). The rhythmic raising and lowering of the floor of the buccopharyngeal cavity, which can be seen easily in a living frog, mixes the air in the buccopharyngeal cavity with the surrounding ambient air. This equalizes the gas content of buccopharyngeal and ambient air. These movements are also important in permitting the frog to smell the air as it is drawn across the olfactory cells in the nasal cavities.

At less frequent intervals there are ventilatory cycles that renew a part of the pulmonary gas, and, from time to time, a series of ventilatory cycles that fully inflate the lungs. Pressure in the lungs is greater than atmospheric pressure because air is pumped into them by a series of active movements of the floor of the buccopharyngeal cavity, after which the glottis is tightly closed.

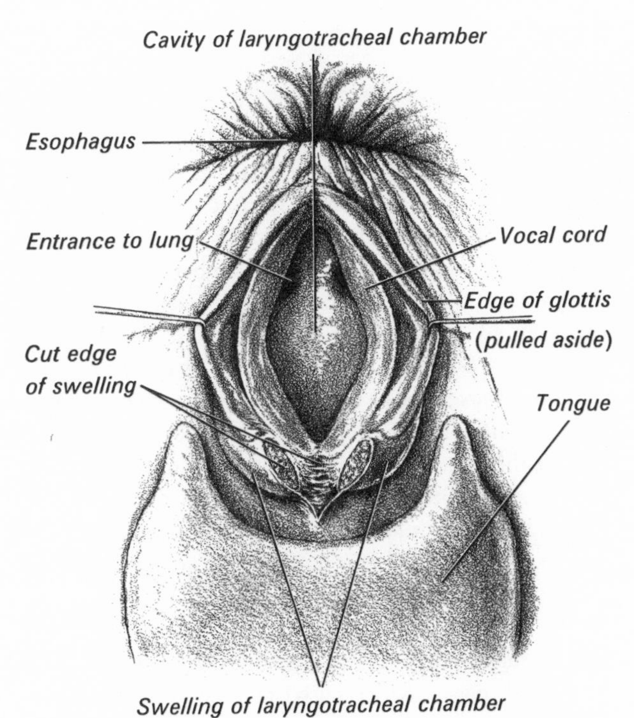

FIGURE 3-7
An enlargement of the floor of the buccopharyngeal cavity of *Rana.* The glottis has been slit, and its lips pulled apart to expose the features within the laryngotracheal chamber.

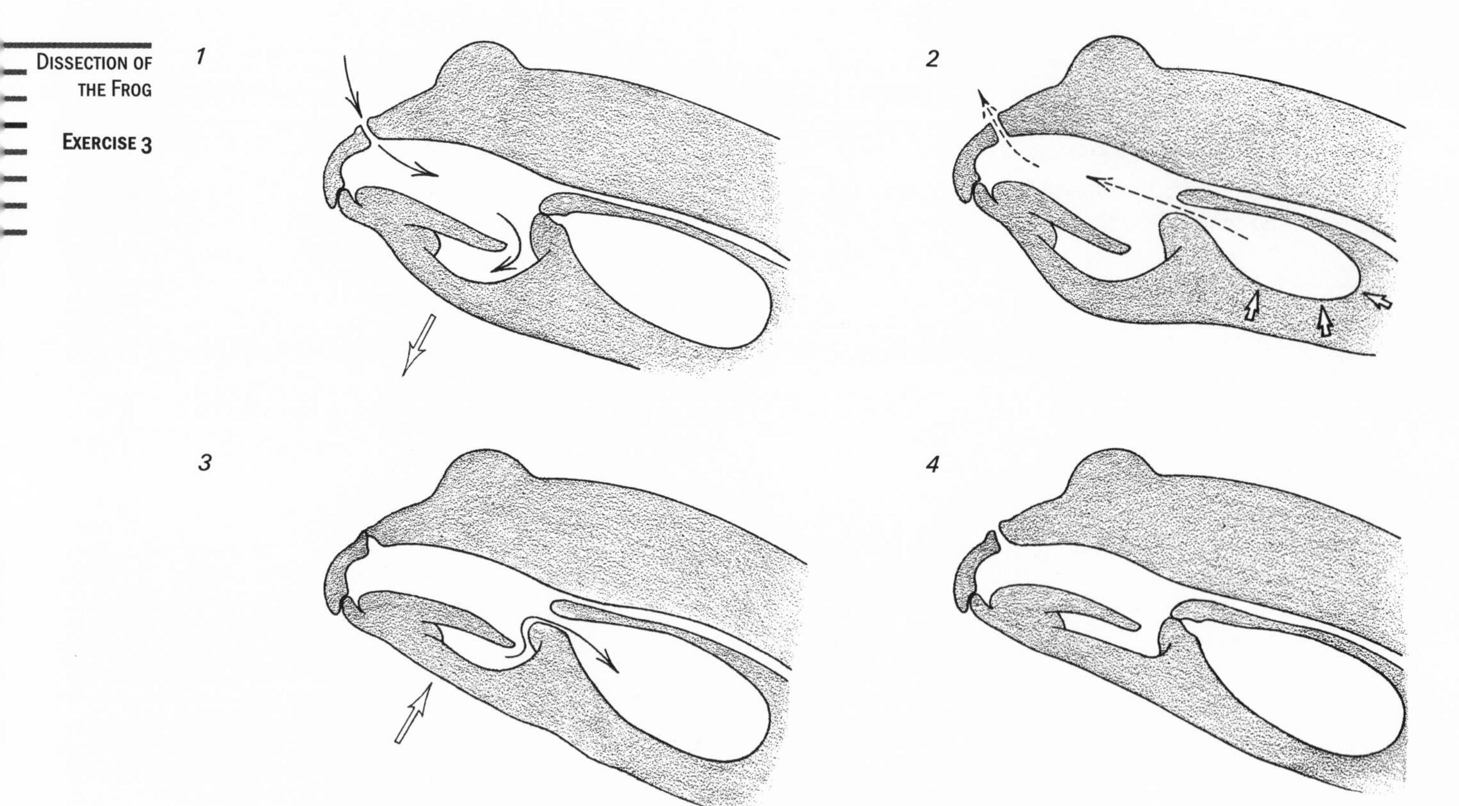

FIGURE 3-8
Diagrams of the four stages in pulmonary ventilation in *Rana*.
Large, open arrows indicate the movements of the floor of the buccopharyngeal cavity and the elastic recoil of the lungs; small, solid arrows, movements of fresh air; broken arrows, movements of spent air. [After de Jongh and Gans.]

Vocalization

In stage 1, a ventilatory cycle begins with a prolonged lowering of the floor of the buccopharyngeal cavity, which reduces pressure in this cavity. Fresh air is drawn through the nasal cavities into a chamber ventral to the tongue and glottis (Fig. 3-8). This is followed in stage 2 by the opening of the glottis and the elastic recoil of the lungs, which had been inflated and stretched in a previous inflation cycle. Some of the depleted air, which has a pressure greater than one atmosphere, flows across the dorsal part of the buccopharyngeal cavity and out of the nostrils. There is little mixing of the exhaled air with the recently inhaled air stored in the ventral part of the buccopharyngeal cavity. As the lungs empty, an inward movement of the flanks is frequently noticed. It is uncertain whether this is a passive movement or an active contraction. In stage 3, the external nostrils close, the floor of the buccopharyngeal cavity is raised, and the air inspired during stage 1 is forced into the lungs, which stretch as they fill. The glottis closes before muscles in the floor of the buccopharyngeal cavity relax (stage 4). During an inflation cycle, ventilation is repeated several times, thereby discharging more depleted air and fully inflating the lungs.

F. VOCALIZATION

Airborne sound waves would be of no value to a highly aquatic animal, and *Xenopus* lacks vocal cords, vocal sacs, and a functional tympanic membrane. Both males and females of other species of frogs have a well developed vocal apparatus including **vocal cords,** a **tympanic membrane,** and **vocal sacs** in the males. Muscles attached to the laryngotracheal chamber can put an appropriate tension on the vocal cords, causing them to vibrate when air is moved back and forth between the buccopharyngeal cavity and lungs. You can usually make a live frog produce small "croaking" noises by stroking the "chest" between the front legs. Male frogs also have vocal sacs, whose entrances you have seen. Expose one of the sacs by removing skin and mucous membrane from the area near the angle of the jaw (Exercise 2, Fig. 2-2). Air can enter the sacs, causing them to swell out and act as resonating chambers. The mating calls of frogs, heard in the spring of the year, are given only by the males. Each species has a distinctive call. Frogs of both sexes can make distress and warning calls.

Circulatory System

THE CIRCULATORY SYSTEM includes the heart, the vessels, and the blood and lymph that the vessels conduct throughout the body. It is a transport system that distributes food, oxygen, and other materials from sites of intake to all of the cells, and that carries carbon dioxide, urea, and other waste products of metabolism to sites of removal. It also carries hormones from the glands in which they are produced to the various target organs. In addition to its transport functions, the blood helps to regulate the amount of water and salts in the interstitial fluid, defends the body against the invasion of disease organisms, and generally helps the organism to maintain a constant internal environment.

A. THE HEART AND ITS ASSOCIATED VESSELS

The heart lies in the pericardial cavity dorsal to the sternum and pectoral girdle. Carefully dissect away the wall of the pericardial cavity to see the chambers of the heart, and the vessels that enter and leave it (Fig. 4-1). The caudal, cone-shaped part of the heart, which is more muscular than the rest, is the **ventricle.** A tubular chamber of the heart, the **conus arteriosus** leaves the right side of the front of the ventricle and carries blood cranially to a pair of large arteries (each is a **truncus arteriosus**), which

The Heart and Its Associated Vessels

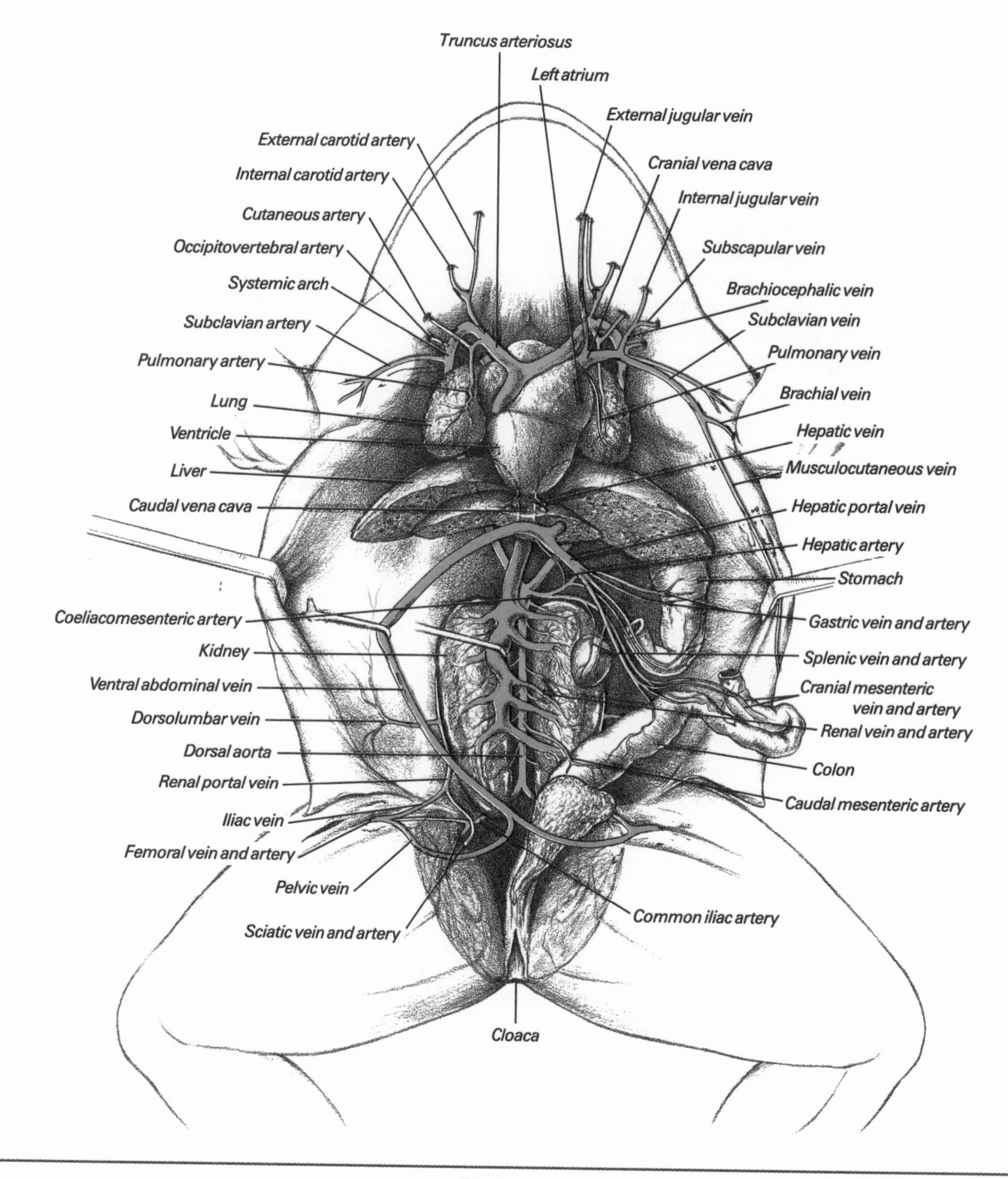

FIGURE 4-1
Semidiagrammatic ventral view of the circulatory system of *Rana*. Many of the viscera have been removed. Cranial veins are shown only on the right side of the drawing. The position of the sinus venosus dorsal to the atria and ventricle is shown by dotted lines. (See Figs. 4-3 and 4-4 for further details.)

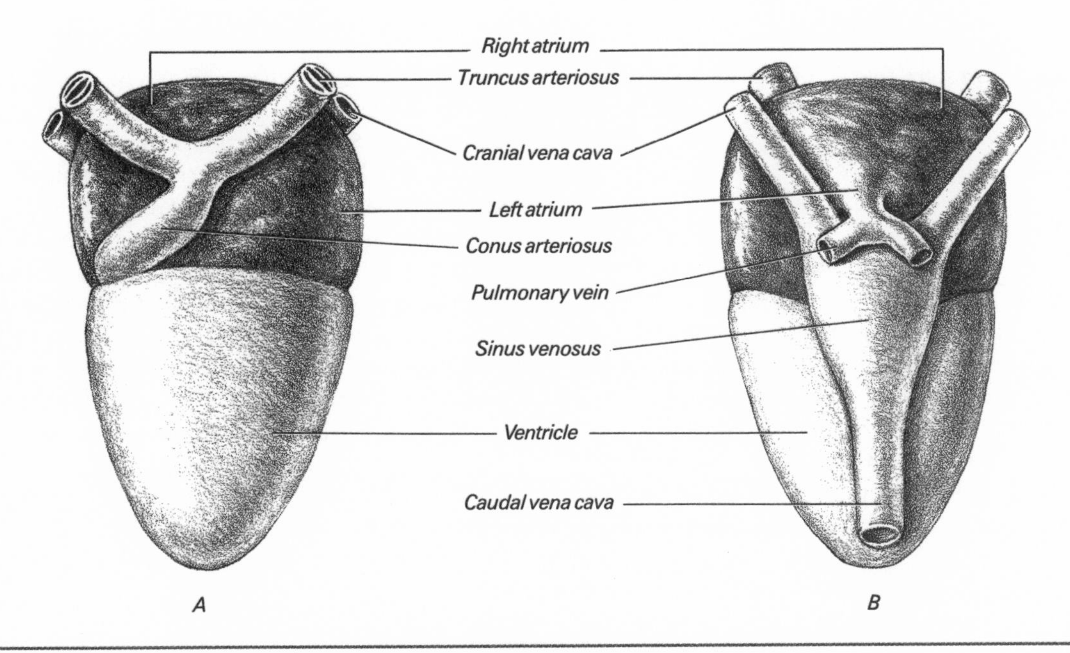

FIGURE 4-5
Ventral (A) and dorsal (B) views of the heart of *Rana*.

tributaries are **gastric veins** from the stomach and **mesenteric veins** from the intestine and spleen.

5. The Renal Portal System and Ventral Abdominal Vein

Trace the ventral **abdominal vein** caudally (Fig. 4-4). It receives tributaries from the body wall and the urinary bladder, and then is formed by the confluence of two **pelvic veins.** Follow one of the pelvic veins. It extends dorsally and laterally and joins a **femoral vein** that comes from the leg. The common trunk thus formed is the **iliac vein.** A **sciatic vein,** which accompanies the corresponding artery in the leg, joins the iliac vein to form the **renal portal vein.** The renal portal vein continues forward along the dorsolateral edge of the kidney. As the name implies, blood in it enters capillaries in the kidney, and these, in turn, are drained by the caudal vena cava. A **dorsolumbar vein,** which drains a large part of the back, also enters the renal portal vein.

This complex arrangement of vessels allows some of the blood from the hind leg to flow forward through the ventral abdominal vein and the liver, and some to flow through the renal portal system and the kidney. The functional significance of a renal portal system, which is present in most nonmammalian vertebrates is not completely understood. The kidney tubules (Exercise 5) of these vertebrates receive blood from both the arterial system and renal portal system. Such a dual supply ensures a full supply of blood to the kidneys, which clear the blood of waste products and surplus materials. A renal portal system may be particularly important in amphibians in which arterial pressures are relatively low.

Heart

D. HEART

Study the beating of the heart on a demonstration dissection of a pithed frog. Notice that the sinus venosus is the "pacemaker." It contracts first, sending blood into the right atrium. Both atria contract next, sending their blood into the single ventricle. Contraction of the ventricle and conus greatly increases the blood pressure and drives the blood to the lungs and body. Valves located between the various chambers prevent a backflow of blood.

Carefully cut across each truncus arteriosus, the cranial and caudal venae cavae, and the pulmonary veins. Then remove the heart from your specimen. Identify again the chambers of the heart and the vessels entering and leaving it as seen both in a ventral and a dorsal view (Fig. 4-5).

The internal structure of the heart is illustrated in Figure 4-6. Some of the structures shown can be seen in a large specimen by carefully cutting away the ventral surfaces of the ventricle, the atria, and the conus arteriosus with a very sharp instrument such as a razor blade. Pick away coagulated blood and the injection mass from within the heart chambers. Notice the spongy **ventricular musculature,** the **interatrial septum** separating left from right atrium, the common opening of the two atria into the ventricle, and a spiral fold, known as the **spiral valve,** in the conus arteriosus. Other internal features will be difficult to see.

E. CIRCULATION

Circulation

The course of blood flow through the frog's heart and body has been a matter of considerable interest and study. Blood from the lungs returns to the left atrium. When the frog is breathing air, this blood will be rich in oxygen but will still contain considerable carbon dioxide. Blood returning to the right atrium will be mostly oxygen depleted blood from the body tissues, but there will also be some blood from the skin that has lost most of its carbon dioxide and gained some oxygen. Since both atria enter a common ventricle, the two atrial bloodstreams can potentially be mixed. Recent studies have shown that although some mixing occurs, the atrial bloodstreams remain largely separate because of the interaction of several mechanisms. For one, the spongy ventricular musculature prevents the two streams from swirling around and mixing freely (Fig. 4-6). In addition, a complex, tripartite division within each truncus arteriosus channels blood to the three aortic arches. Next, a deflective action of the spiral valve separates blood going to the pulmocutaneous arch from that going to other arches. Differences in peripheral resistance make

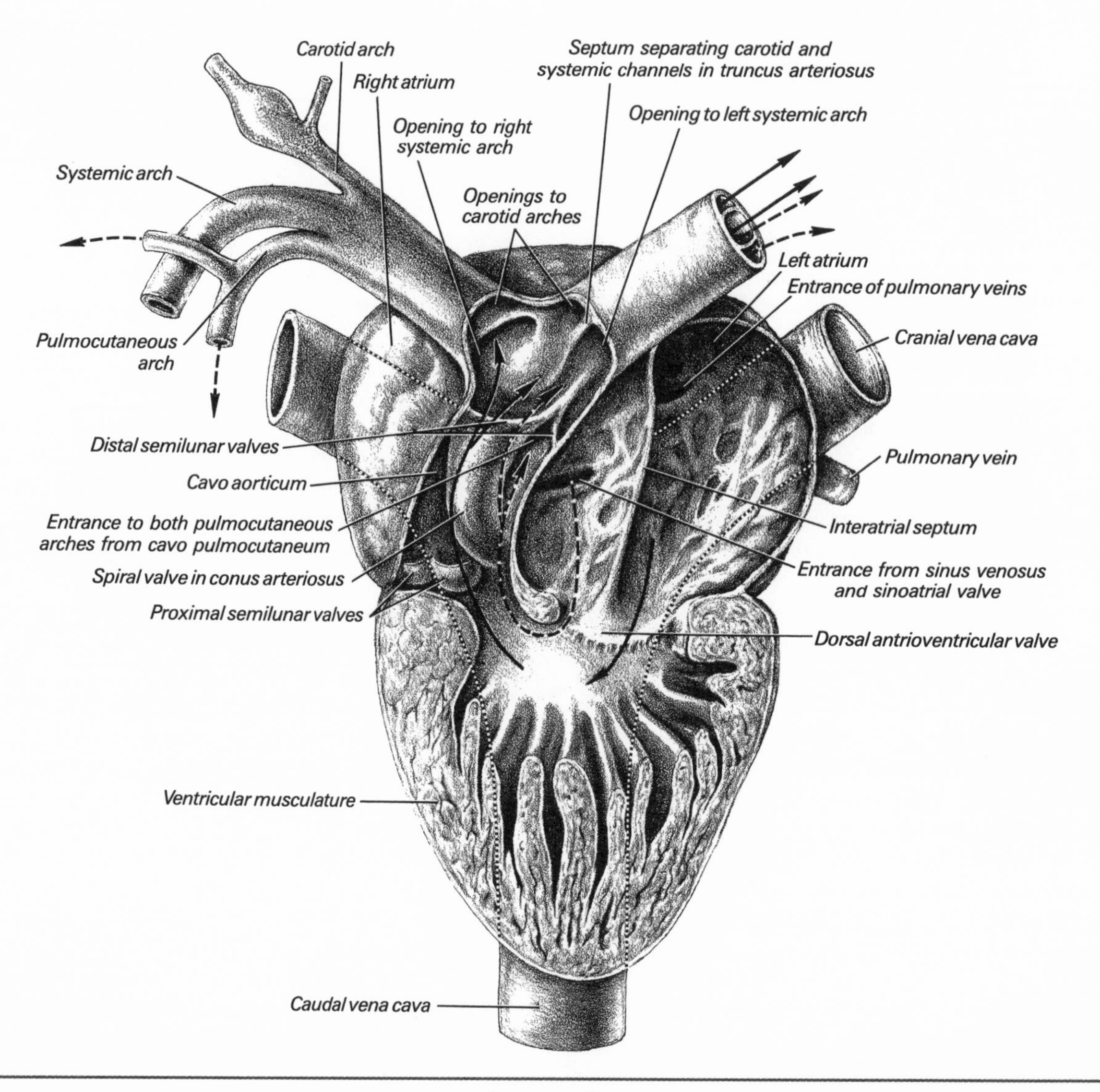

FIGURE 4-6
Ventral view of the heart, partly sectioned in the frontal plane to show internal structure. Broken lines show the position of the sinus venosus on the dorsal surface. Arrows indicate the distribution of blood from the two atria. (See text for discussion.)

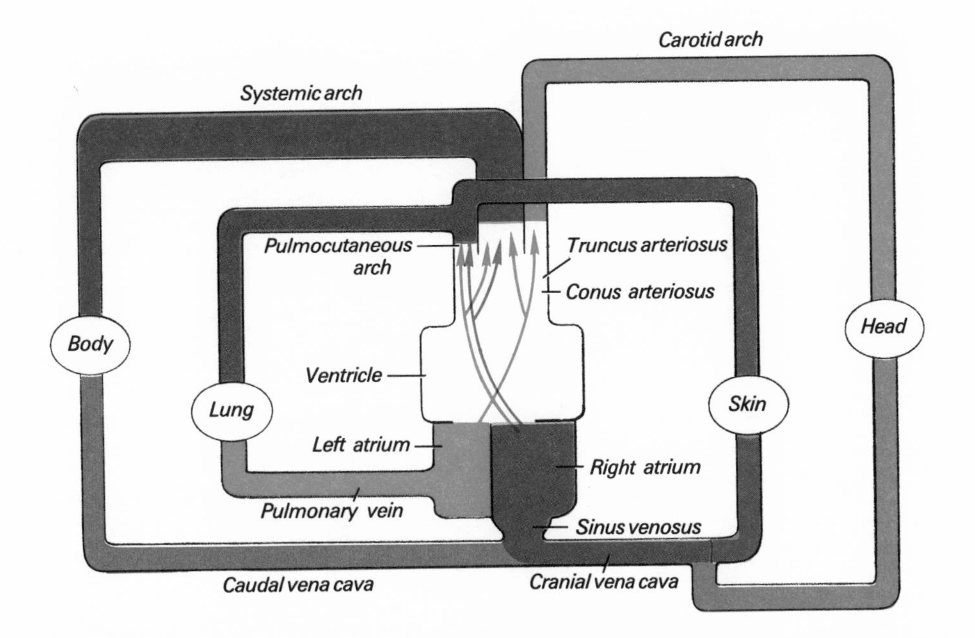

FIGURE 4-7
Schematic diagram of the pattern of circulation in a ranid frog. The relative oxygen content of blood in the vessels is shown by the degree to which the vessels are colored red (high oxygen content) or blue (low oxygen content).

it easier for blood to go to some arches than to others. Finally, the volume of blood returning to each atrium can affect the pattern of distribution.

As a result of the interaction of these factors (Fig. 4-7), the carotid arches carry primarily oxygen-rich blood from the left atrium to the head. The pulmocutaneous arches receive primarily right atrial blood, which is low in oxygen content, having only as much as is returned from the skin. Blood in the systemic arches going to the body is mixed, but contains more oxygen-rich left atrial blood than depleted right atrial blood. Some evidence suggests that, in ranid frogs, the blood in the left systemic arch is mixed, whereas that in the right systemic arch is oxygen-rich.

A probable advantage of this complex arrangement is that the degree to which blood from the two atria is mixed can be varied according to the needs of the frog and the amount of pulmonary and cutaneous gas exchange that is taking place. When a ranid frog is under water for a considerable time, as it is during hibernation, all of its gas exchange occurs through the skin. Pulmonary resistance to blood flow presumably would increase, as it does in other terrestrial vertebrates when they are beneath water, and less blood would be sent through the functionless lungs. All aortic arches would receive a mixed blood. *Xenopus*, as you have seen, has surprisingly large lungs, and blood flow through them is greater than in *Rana*. The increased volume of blood returning to the left atrium changes balance within the heart in such a way that oxygen-rich blood goes to the carotid arches and to both systemic arches.

Circulation

Most fishes have a heart with a single atrium. Venous blood low in oxygen content is received from the body by the sinus venosus. It passes through each chamber in turn, and flows out to the gills where it is aerated. This arterial blood continues on to the body. Because mammals have a heart with two atria and two ventricles, the two types of blood going through the heart can be kept completely separate. Oxygen depleted blood is received from the body by the right side of the heart and is sent to the lungs. Oxygen rich blood returns from the lungs to the left side and is sent to the body. It was once assumed that the partial separation of bloodstreams in the frog heart is the necessary intermediate evolutionary stage between the single bloodstream in the fish heart and the complete separation of two bloodstreams in the mammalian heart. However, the interventricular septum and the nearly complete separation of the two bloodstreams in lungfish may reflect the ancestral amphibian condition. The absence of an interventricular septum in contemporary frogs, and the consequent possibility of some blood mixing, is now regarded as a corollary of cutaneous respiration, which has evolved in modern amphibians.

Urogenital System

THE KIDNEYS ARE EXCRETORY ORGANS that eliminate nitrogenous waste products of protein metabolism, but they are not solely excretory organs. The kidneys eliminate a variety of materials that may be present in the body fluids in amounts exceeding the body's needs, and they conserve those materials not in excess. In so doing, the kidneys play an essential role in maintaining an internal environment that is nearly constant in water and salt content, in pH, and in the blood level of sugar and many other substances. The reproductive organs obviously have a very different function, but it is convenient to consider certain aspects of these two systems together, because the reproductive cells of the male are discharged through urinary ducts.

A. THE EXCRETORY SYSTEM

The urogenital organs can be seen by pushing the stomach and intestines aside, but you may remove much of the digestive system if you wish. Do not remove the caudal part of the large intestine, the base of the coeliac artery, or the cranial part of the liver and lungs.

Each **kidney** is an elongated oval organ located against the back. Parietal peritoneum covers its ventral surface, and its dorsal surface bulges into the subvertebral lymph sac (Exercise 3, Fig. 3-5). The kidneys are composed of several thousand **kidney tubules,** which produce the urine. It will be recalled from Exercise 4 that arterial blood is brought to each kidney by urogenital arteries, and that venous blood is carried to it by the renal portal veins. The interrelationships of these two blood supplies to a tubule is show diagrammatically in the insert to Figure 5-1. Notice that only arterial blood goes to a knot of capillaries, the

The Excretory System

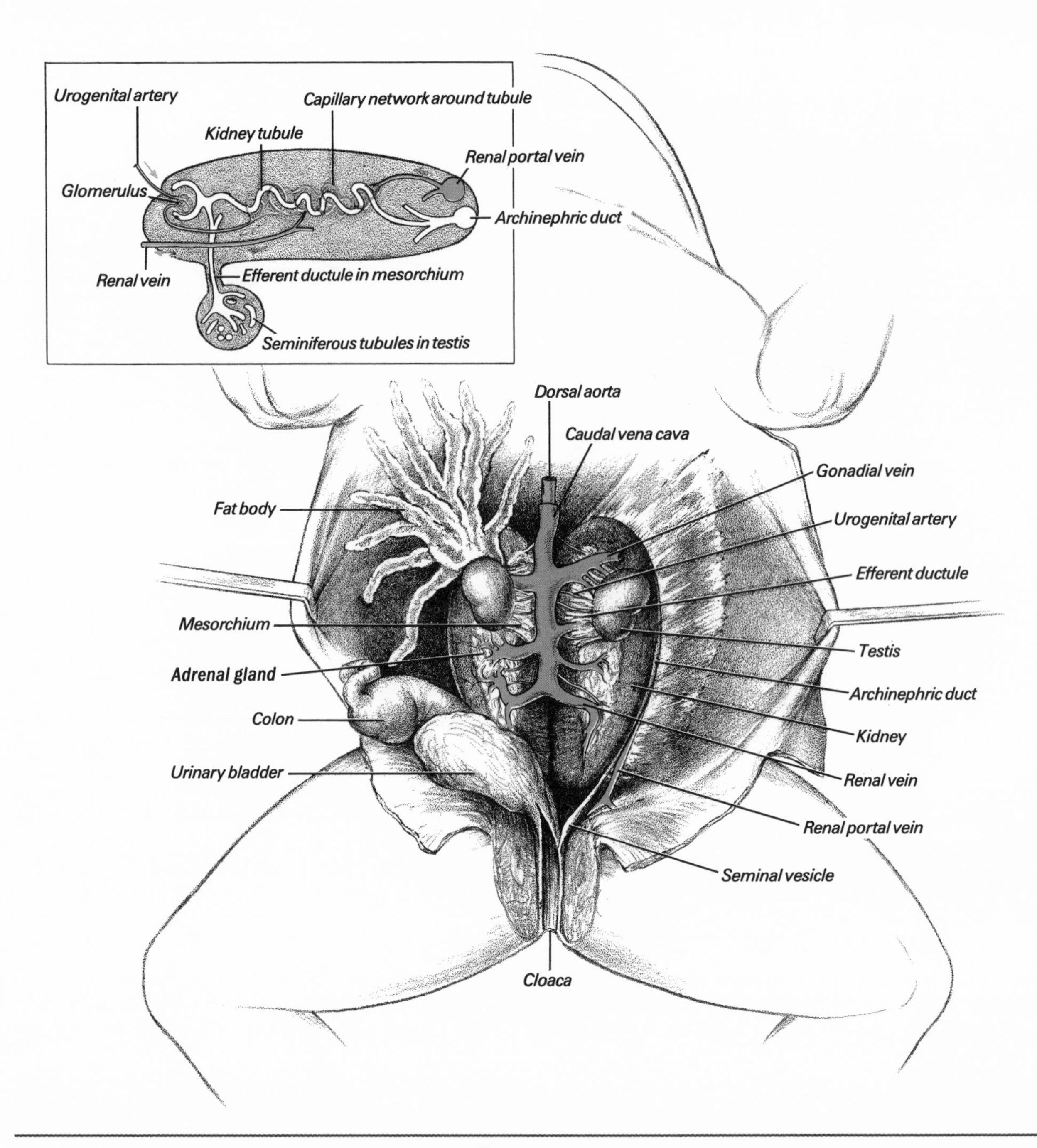

Figure 5-1
Ventral view of the urogenital system of a male *Rana.* One fat body has been removed. The insert is a diagrammatic cross section through a kidney and testis.

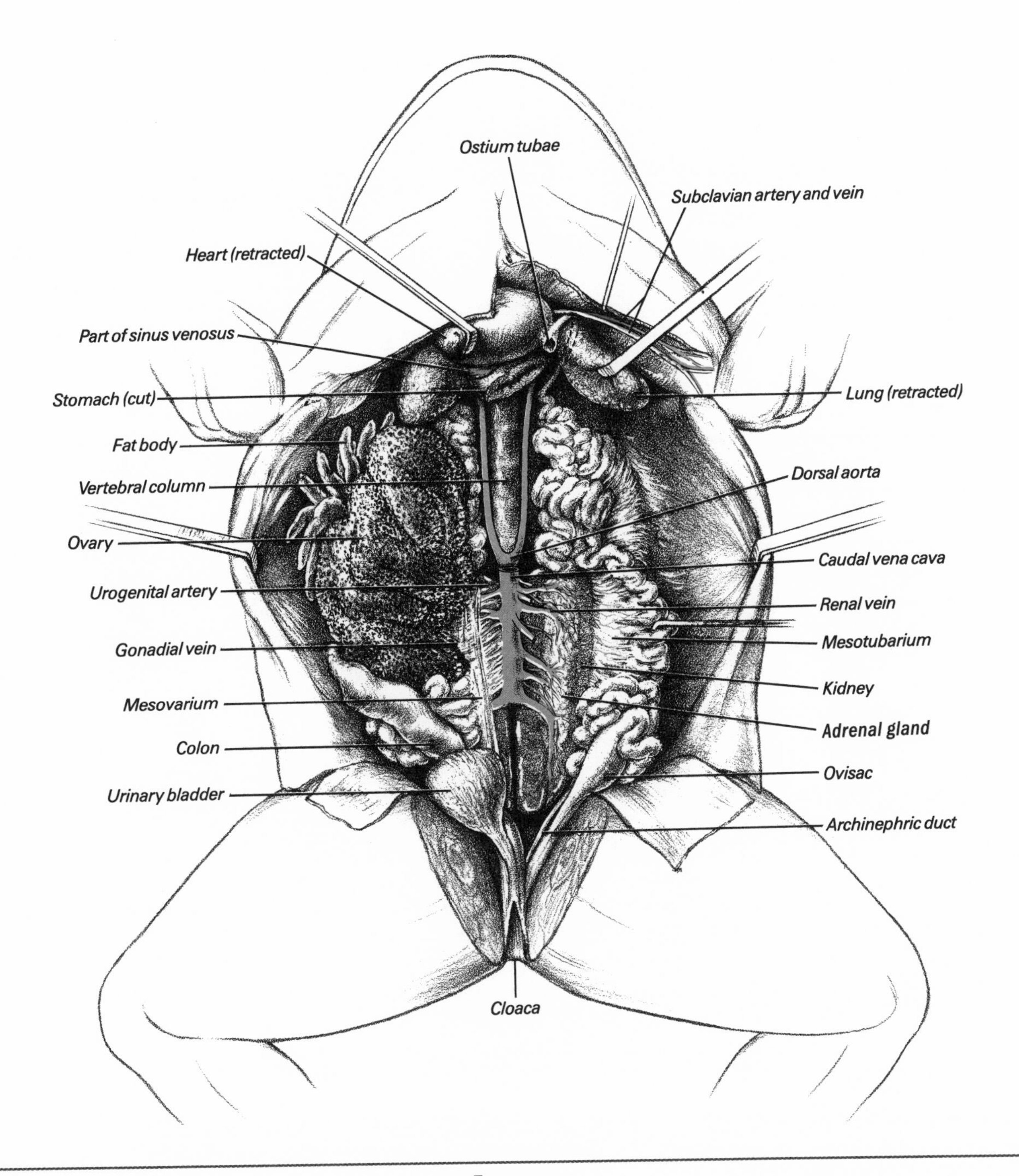

FIGURE 5-2
Ventral view of the female urogenital system of *Rana*. One ovary and one fat body have been removed. Because this drawing is based on a specimen taken in midsummer, the ovaries and fat bodies have not reached their full size.

glomerulus, that has pushed into the **glomerular capsule,** Bowman's capsule, at the beginning of a tubule. Filtration here removes much water, salts, nitrogenous wastes, and other small molecules from the blood, but not plasma proteins and large molecules which are held back. As the filtrate continues down the tubule, salts, sugars, water, and other substances needed by the body are selectively reabsorbed into capillary beds surrounding the tubule. Additional substances, among them nitrogenous wastes, are also secreted into the tubule in this region. The filtrate acquires a progressively greater concentration of waste products, and the final urine is formed. Capillary beds around the tubules are drained by the renal veins, which lead to the caudal vena cava.

If you have not done so previously (Exercise 3), cut through the ventral surface of the pelvic girdle and trace the large intestine and **urinary bladder** into the pelvic canal. They join at the **cloaca.** The urine from each kidney is drained by an **archinephric,** or **Wolffian, duct,** which can be seen along its lateral edge (Figs. 5-1 and 5-2) lying ventral to the renal portal vein. If an oviduct is present, as it is in females and some males, the archinephric duct lies along the line on which the mesentery supporting the oviduct is attached to the kidney. The duct is larger in males than in females because it also carries sperm. Trace an archinephric duct caudally and find the site at which it joins the dorsal surface of the cloaca. Because the urinary bladder attaches to the ventral surface of the cloaca, urine must flow across the cloaca to enter the bladder.

In their normal amphibious environment, frogs must eliminate, along with nitrogenous wastes, a great deal of water, which has diffused into the body, yet they must conserve body salts. Production of a copious and dilute urine is therefore adaptive. Much of the nitrogenous wastes is eliminated as urea, but the high flow of water into and out of the body permits much to be eliminated as ammonia. **Ammonia** is one of the simplest forms in which nitrogen can be eliminated, but because it is very toxic it must be carried off in low concentrations by a large quantity of water. **Urea** is biochemically a more complex product, but it is less toxic and requires less water for its removal. Ammonia production predominates in larval amphibians and in such highly aquatic species as *Xenopus*. Enzymes needed to synthesize urea develop at metamorphosis, partly under the influence of thyroxine, and adult amphibians excrete much more urea than ammonia.

The Reproductive System

Frogs have less ability to control water loss under conditions of dehydration than do truly terrestrial vertebrates, but certain countermeasures can be taken. The rate of glomerular filtration can be decreased, and the rate of water reabsorption from the kidney tubule can be increased. These two processes together result in a decreased volume of urine and in the production of a more concentrated urine. Also, some water can be absorbed from the urinary bladder. Indeed, certain tropical frogs can store water in the bladder during the rainy season for use during the dry season. Most frogs cannot withstand dehydration for such a long period of time. They must return frequently to the water, where they can take up through the skin water that has been lost. Salts can also be absorbed through the skin.

B. THE ADRENAL GLANDS

Notice the irregularly shaped band of light tissue lying on the ventral surface of each kidney. This is the **adrenal,** or **suprarenal, gland.** Cells in its central, or medullary, region secrete **epinephrine,** or **adrenaline,** a hormone that increases blood sugar levels, increases the rate of heart beat, and in general helps the body to adjust to conditions of stress. Its outer, or cortical, cells produce a variety of **cortical hormones,** which affect carbohydrate metabolism, mineral metabolism, and the water and salt balances of the body. In adrenalectomized frogs, the normal exchanges of water and salts through the skin are upset.

C. THE REPRODUCTIVE SYSTEM

Reproduction is seasonal in temperate species of frogs, but those living in the tropics reproduce several times a year.

Dissect the reproductive organs of your specimen very carefully to show all parts clearly. After studying your specimen, share it with a fellow student who has dissected one of the opposite sex.

1. The Male Reproductive Organs

Each **testis** is a small oval organ attached by a mesentery, the **mesorchium,** to the ventral surface of a kidney (Fig. 5-1).

A **fat body,** which consists of long fingerlike lobes of tissue, lies at the cranial end of the testis. In North American and European ranid frogs, fat bodies are largest during the fall of the year. Some of the food stored in them is used during hibernation, and the rest is rapidly depleted during the spring breeding season, when the frogs become very active.

The testes are composed of microscopic **seminiferous tubules** in which spermatogenesis occurs. Testes of temperate species of frog enlarge during the late winter and spring breeding season. Sperm leaving the seminiferous tubules first pass through a microscopic, **longitudinal duct of the testes,** and then enter minute **efferent ductules,** which extend through the mesorchium to the kidney. If your specimen is well injected, these ductules can be distinguished from the arteries and veins supplying the testis: the ductules are the uninjected passages. Within the kidney, the ductules branch, and each one connects with the proximal ends of eight to twelve of the cranial kidney tubules. These kidney tubules convey the sperm to the **archinephric duct.** It is not clear whether urine function is modified in these tubules during sperm transport. Sperm are carried down the archinephric duct to the **seminal vesicle,** which is a slight dilation on the caudal end of the duct. Sperm are stored here temporarily until the male frog mounts the back of the female, an embrace called **amplexus.** Sperm are discharged through the male's **cloaca** as the eggs are laid.

Groups of interstitial cells are interspersed among the seminiferous tubules. They secrete the male sex hormone that influences male breeding behavior and the development of secondary sex characters. Breeding males of *Rana* have enlarged thumbs (Exercise 1) that help them grasp the females; those of *Xenopus* have dark, horny tubercles on their fingers and forearms.

During development, the embryos of frogs and other vertebrates pass through a sexually indifferent stage, during which the primordia of both male and female reproductive ducts are present and the embryo has the morphological possibility of developing into either sex. Normally one set of passages continues to develop and those of the other set atrophy, but in some adult male ranid frogs a rudimentary oviduct may be found lateral to each kidney.

2. The Female Reproductive Organs

Each **ovary** is a multilobed organ attached by a mesentery, the **mesovarium,** to the ventral surface of a kidney (Fig. 5-2). The ovaries vary greatly in size according to the season. If they are very large, remove one ovary in order to see other organs more clearly. In Northern seasonal breeding ranid frogs, the ovaries are smallest after the frogs reproduce in the spring. Unlike those of many other female vertebrates, the primordial egg cells, or **oogonia,** of frogs can multiply throughout the reproductive life. During the summer a crop of several thousand oogonia begin to enlarge and mature and accumulate yolk. Yolk deposition in the egg cells continues during winter hibernation, using lipid materials stored in the **fat bodies.** Fat bodies appear as fingerlike lobes of yellowish tissue attached to the cranial border of the mesovaria.

The developing eggs are surrounded by a thin layer of follicle cells, which can be seen only microscopically. As in other female vertebrates, the follicular cells produce the female sex hormones under the stimulation of pituitary hormones. When a certain hormonal balance is attained, the eggs are **ovulated;** that is, they rupture the wall of the follicle and ovary and are discharged into the pleuroperitoneal cavity. At the time of ovulation, the eggs are in the prophase of their first meiotic division. This division is completed rapidly after ovulation.

Amphibian ovulation can also be triggered by other hormones, including human chorionic gonadotropin, which is produced by the placenta in very early stages of pregnancy and appears in the urine. Pregnancy tests are based on its detection. A few years ago, assays were made by injecting urine extracts into female *Xenopus.* If chorionic gonadotropin was present, *Xenopus* would ovulate in a few days.

A long, convoluted **oviduct** lies against the back on each side of the body. Trace one forward. At the cranial end of the pleuroperitoneal cavity it passes lateral to the base of the lung. In *Xenopus* it ends here, but in *Rana* it curves ventrally beside the base of the lung for a short distance (Fig. 5-3). In all vertebrates it terminates in a funnel-shaped opening, the **ostium tubae.**

At the time of ovulation, a large part of the coelomic epithelium is ciliated, and ciliary currents carry the eggs toward the ostium. Cilia in the oviduct carry the eggs into the ostium, and cilia and peristaltic contractions of oviductal muscles propel them down the oviduct. As the eggs move down the oviduct, the second meiotic division is initiated, and certain oviductal cells secrete around the eggs layers of jelly that consists largely of albumin and muco-

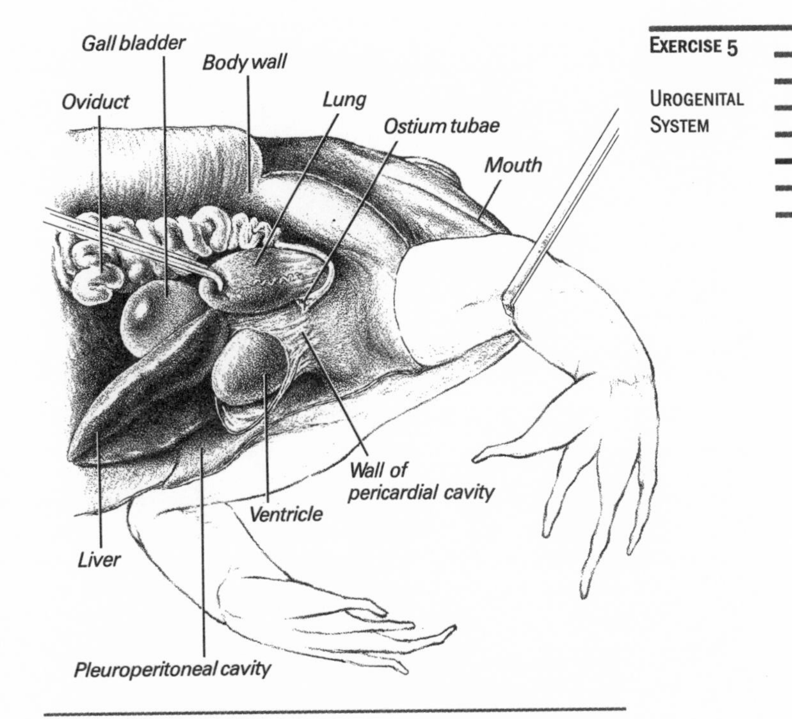

FIGURE 5-3
The relationship of the cranial portion of the oviduct to the surrounding organs is shown in this lateral view of the right side of *Rana.*

The Reproductive System

protein. Jelly secretion is under the control of ovarian hormones. The eggs may be stored for a day or two in a thin-walled expansion of the caudal end of the oviduct known as the **ovisac** (Fig. 5-2). The ovisac enters the dorsal surface of the **cloaca** beside the entrance of the archinephric duct. The stimulus of the sexual embrace of the male **(amplexus)** is required for the eggs to be discharged from the ovisac into the cloaca. As the eggs emerge through the cloacal aperture, sperm penetrates them and the layers of jelly imbibe water and swell. The second maturation division of an egg is completed after sperm penetration, but before the union of the male and female nuclei.

The jelly layers help to protect the developing embryo from mechanical injury and from fungus and other infections, and they help to conserve the heat produced by the metabolism of the embryo. After six days of embryonic development, leopard frog embryos hatch and emerge from the jelly envelopes as small tadpoles. These feed for themselves for nearly three months before metamorphosing to adult frogs. Bullfrogs have a longer period of development, usually remaining in the larval stage for two or three years before metamorphosis. *Xenopus* eggs hatch in 50 hours into small tadpoles that grow rapidly and metamorphose at about 58 days of age. The rapidity of *Xenopus* development and the ease of caring for them in the laboratory has made the species a very useful experimental organism.

Nervous System

NO ORGANISM CAN LIVE unless its various parts work together, and unless the organism as a whole can detect relevant changes in its environment and respond to them in a way that promotes its survival. Various receptive cells detect changes in both the internal and the external environments. Muscles, glands, ciliated cells, and chromatophores bring about the response of the organism, whether it be a flick of the tongue, an increase in blood pressure, the secretion of pancreatic juice, or a change in color. Between receptors and effectors lies a complex integrating mechanism, which receives and stores information from the receptors, compares current information with past experience, and activates the appropriate effectors. Rapid and specific integration of the activity of the internal organs, as well as most of the animal's responses to the external environment, are mediated through the nervous system, which is composed of elongated nerve cells, or **neurons.** The specificity of nervous integration is made possible by specificities within the system. From specific receptor cells that are attuned to specific environmental parameters (light, temperature, chemical change, mechanical displacement) neurons extend to specific parts of the spinal cord and brain, and from the brain and cord to appropriate effectors.

Metabolic activities of the body are in large part regulated by the hormones secreted by the endocrine glands. Small quantities of hormones are carried by the bloodstream to all parts of the body, where they affect cells that have the necessary biochemical receptor mechanisms. Some hormones have wide-ranging effects, whereas others have very specific target organs. In previous exercises, the thyroid gland, islets of Langerhans, and the adrenal gland have been mentioned. In this exercise, the pineal gland and hypophysis will be seen.

The neurons are grouped in such a way that it is convenient to divide the nervous system grossly into central and peripheral parts. The **central nervous system** comprises the brain and spinal cord. It is in these organs, and especially in the brain, that the essential integrations take place. The **peripheral nervous system** includes the cranial and spinal nerves and the autonomic nervous system. Cranial and spinal nerves carry sensory impulses from the receptors to the central nervous system, and motor impulses from the central nervous system to the appropriate effectors, chiefly skeletal muscles. Autonomic fibers leave the central nervous system in certain cranial and spinal nerves, and then branch off from these nerves to supply glands and muscles in the walls of blood vessels and visceral organs.

A. THE SPINAL NERVES

Study the peripheral distribution of the spinal nerves before dissecting the spinal cord. Remove all of the abdominal viscera except for the systemic arches, the dorsal aorta, and the base of the coeliacomesenteric artery (Fig. 6-1). Identify the individual vertebrae and their transverse processes. **Spinal nerves** leave the spinal cord and emerge between the vertebrae.

Each spinal nerve divides into three branches: (1) an inconspicuous **dorsal branch** supplying muscles and skin on the back; (2) a prominent **ventral branch** which is the one seen in the spaces between the transverse processes; (3) a small **communicating branch** leading to the sympathetic cord. There are ten pairs of spinal nerves. The origins of their ventral branches are surrounded in *Rana* by **paravertebral lime sacs** consisting of calcareous granules that are deposited in extensions of the endolymphatic sacs of the

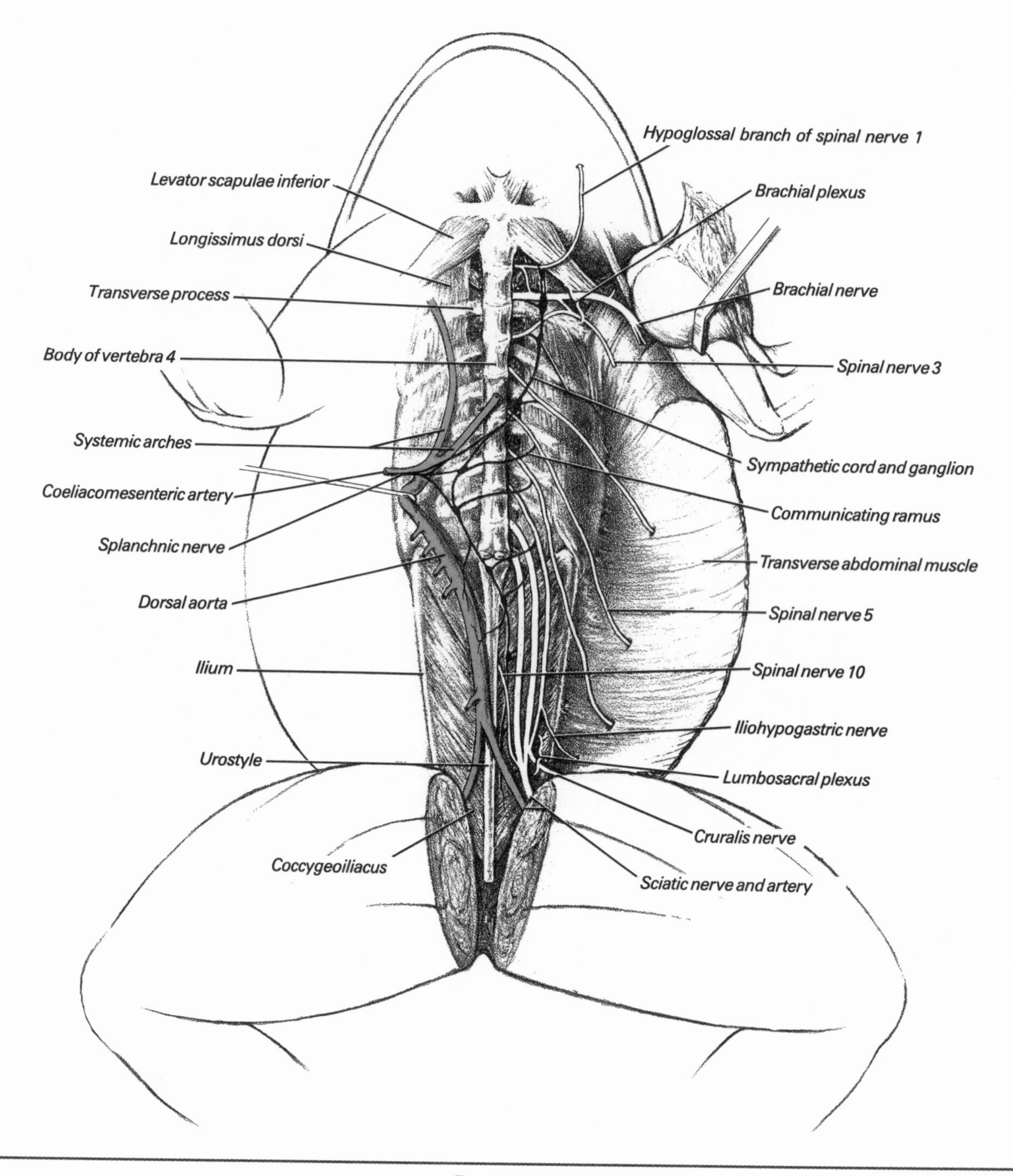

FIGURE 6-1
Ventral view of the spinal nerves and the sympathetic cord of *Rana*.

inner ear. These paravertebral lime sacs are calcium reserves for bone formation and many other biochemical processes. Such sacs are not well developed in *Xenopus*.

The ventral branch of the most cranial spinal nerve can be seen emerging from between the first and second vertebrae. Its major part, which extends both ventrally and cranially, supplies primarily muscles in the floor of the buccopharyngeal cavity and in the tongue. These fibers are comparable to ones in the hypoglossal nerve of mammals, which is a cranial nerve that supplies similar muscles. A small branch of this spinal nerve extends caudally to join the next nerve. Because of the location of this spinal nerve in adult amphibians, it is usually called the first spinal nerve. However, it is homologous to the second spinal nerve of other vertebrates. The true first spinal nerve of vertebrates emerges between the skull and the first vertebra. Such a nerve is present in early larval frogs but is lost later in development.

The second spinal nerve of the frog is the largest of all. It, together with small branches it receives from the first and third nerves, forms a network known as the **brachial plexus.** Nerves from the plexus supply the muscles and skin of the shoulder and arm. The main continuation of the second nerve into the arm is the **brachial nerve.** After contributing to the brachial plexus, the third spinal nerve, along with the fourth, fifth and sixth nerves, supplies much of the body wall.

The seventh, eighth, and ninth spinal nerves form the **lumbosacral plexus,** from which nerves extend to the muscles and skin of the caudal part of the abdomen, pelvis, and hind leg. The largest of these nerves is the **sciatic nerve,** which accompanies the sciatic artery and vein into the thigh.

The tenth spinal nerve is a very small nerve, which emerges from a foramen on the side of the urostyle. It supplies the cloaca, the urinary bladder, and other organs in this region. In some individual frogs, it gives off a small branch that joins the lumbosacral plexus, and in some it receives a small branch from the ninth nerve.

B. THE SYMPATHETIC CORD

The **autonomic nervous system** carries motor fibers to glands, blood vessels, and other visceral organs. It can be divided into **sympathetic** and **parasympathetic** portions; neurons of the sympathetic portion leave the central nervous system through spinal nerves, and those of the parasympathetic portion leave through certain cranial nerves. Most visceral organs are supplied by both portions, which have antagonistic effects upon a given organ. Sympathetic stimulation, along with the secretion of the hormone adrenaline by medullay cells of the adrenal gland, help the body adjust to such stresses as escaping a predator. Sympathetic stimulation and adrenalin mobilize mechanisms that provide more energy to skeletal muscles: heat rate increases, blood sugar levels increase, blood pressure rises and more blood is delivered to skeletal muscles. Digestive processes are inhibited. Parasympathetic stimulation tends to have the opposite effects and restores energy reserves.

Another unique feature of the autonomic system is that motor neurons beginning in the central nervous system do not extend all the way to the effectors as they do in other peripheral nerves. Rather, they terminate in **peripheral ganglia,** where they synapse with the dendrites and cell bodies of a second group of neurons that continue to the effectors.

The most conspicuous parts of the autonomic system are the two **sympathetic cords,** one on each side of the vertebral column ventral to the transverse processes (Fig. 6-1). On close examination you will see one or more tiny **communicating branches** leading from each spinal nerve to a sympathetic cord. The slight enlargements on the cord at the sites where the communicating branches join are the **sympathetic ganglia.** It is in them that many of the abovementioned relays between preganglionic and postganglionic neurons occur. Small nerves leave the cord and follow the major arteries to the effector organs. The largest of these nerves is the **splanchnic nerve,** which follows the coeliacomesenteric artery. Sensory fibers returning from receptors in the viscera lie in the same nerves that carry motor fibers, but, as a matter of definition, they are not considered to be a part of the autonomic nervous system.

The Spinal Cord and the Origin of the Spinal Nerves

C. THE SPINAL CORD AND THE ORIGIN OF THE SPINAL NERVES

Skin the back, cut through the tough fascia covering the back muscles, and remove enough muscle to expose completely the top of the vertebral column and the back of the skull. You will have to cut away the dorsal part of the pectoral girdle (suprascapula and scapula, Exercise 1). Bend the head ventrally and carefully cut open the joint between the skull and the first vertebra. Look for the **spinal cord** lying in the **vertebral canal.** Carefully insert one point of a pair of scissors into the canal and cut through the vertebral arch of one vertebra. Repeat this on the other side and lift off the top of the vertebra. Continue this procedure for all of the other vertebrae. The spinal cord is surrounded by a nonvascular layer of connective tissue, the **dura mater.** Leave it in place, because it will help protect the cord from injury.

Identify the spinal nerves and observe their departure from the vertebral canal through the **intervertebral foramina,** which lie between successive vertebrae (Fig. 6-2). Notice that the spinal cord is not uniform in diameter; there are slight enlargements adjacent to the origins of those nerves going to the front and hind limbs. These are the **brachial** and **lumbar enlargements.** They contain the

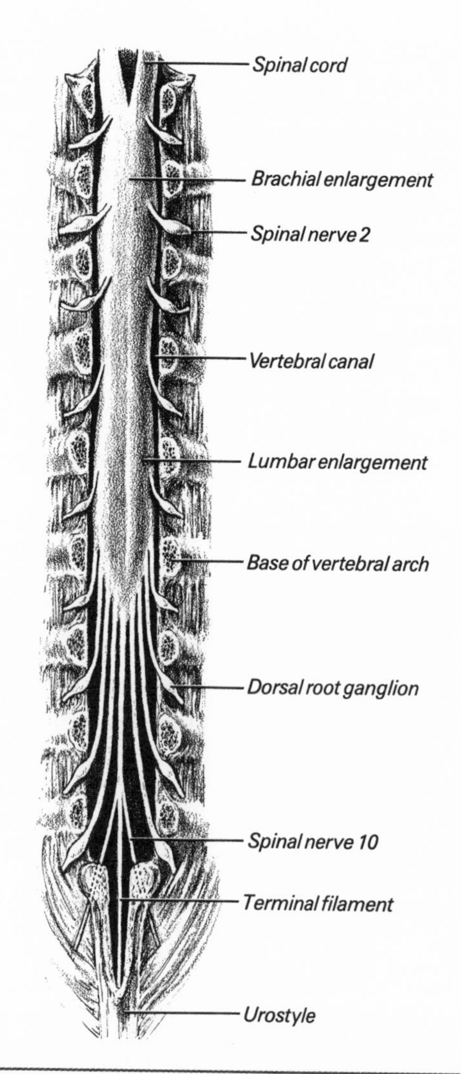

Figure 6-2
Dorsal view of the spinal cord and the origins of the spinal nerves.

cell bodies for the greater number of motor neurons going to the limb muscles. The cord ends in a fine **terminal filament** extending into the urostyle.

Slice open the dura mater and carefully cut away more bone and muscle. Follow the large second nerve laterally. If possible, use a dissecting microscope. You can see that this nerve, like the others, arises from the spinal cord by **dorsal** and **ventral roots** (Fig. 6-3). Sensory and motor neurons in the spinal nerve segregate in these roots; sensory neurons enter the cord through the dorsal root, and motor neurons leave through the ventral root. After traversing the intervertebral foramen, the roots unite to form the spinal nerve. Just before their union, the dorsal root bears an enlargement, the **spinal ganglion,** which contains the cell bodies of the sensory neurons. If you are fortunate, you may see the small **dorsal branch** of the spinal nerve extending to the back. The largest part of the nerve is the **ventral branch.** Finally, there is a small **communicating branch** to the sympathetic cord that was previously seen in a ventral view (Fig. 6-1).

D. THE BRAIN

The Brain

1. Dorsal Dissection of the Brain

Skin the top of the head in the middorsal region, and remove the muscles from the caudal region of the skull. If the sense organs are to be studied (Exercise 7), do not disturb them on at least one side of the head. From a point midway between the eyes, the brain extends caudally to the back of the skull (Exercise 7, Fig. 7-1). Carefully chip away enough bone to expose it completely. The brain is loosely surrounded by a nonvascular layer of connective tissue, the **dura mater,** and is tightly invested by a more delicate vascular **pia mater.** These membranes, or **meninges,** cover the spinal cord as well, although only the dura was observed. They must be picked away to see the parts of the brain clearly.

The rostral end of the brain is made up of a pair of **olfactory bulbs,** which receive the pair of large **olfactory nerves** extending caudally from the nose (Fig. 6-4A). A pair of larger **cerebral hemispheres** lie caudal to them. Olfactory bulbs and cerebral hemispheres together constitute a brain region known as the **telencephalon.** This is the first of five regions into which the brain is divided. Each region develops from a distinct enlargement of the neural tube of the embryo.

The second brain region is the somewhat low, or depressed, **diencephalon,** which lies caudal to the cerebral hemispheres. A tiny **pineal gland,** which is connected by a nerve tract to the **frontal organ** on the top of the head (Exercise 1), arises from the roof of the diencephalon. The

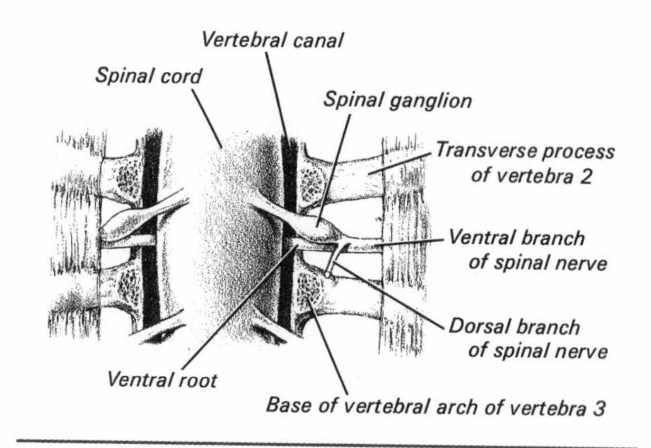

Figure 6-3
Enlargement of the origin of the second spinal nerve; it has been deeply dissected on the right side.

frontal organ contains photoreceptive cells that help the frog adjust physiologically to the diurnal light-dark cycle. The pineal gland contains secretory cells that produce **melatonin,** a hormone that causes pigment in the chromatophores in the skin to aggregate and the animal to blanch. Light received by the frontal organ and eyes can suppress this reaction (by inhibition of melatonin release), and the animal remains dark. In the absence of melatonin, the chromatophores are under the influence only of a **melanocyte stimulating hormone** (MSH) produced by the hypophysis (described below). The MSH causes the pigment to be dispersed.

That part of the roof of the diencephalon that lies cranial to the attachment of the pineal gland is quite thin and vascular. As you pick off the roof, you may notice small, vascular folds, the **choroid plexus,** hanging from it and dipping into the **third ventricle,** one of the cavities within the brain. The walls of the diencephalon, which are lateral to the third ventricle, are relatively thick and constitute the **thalamus.**

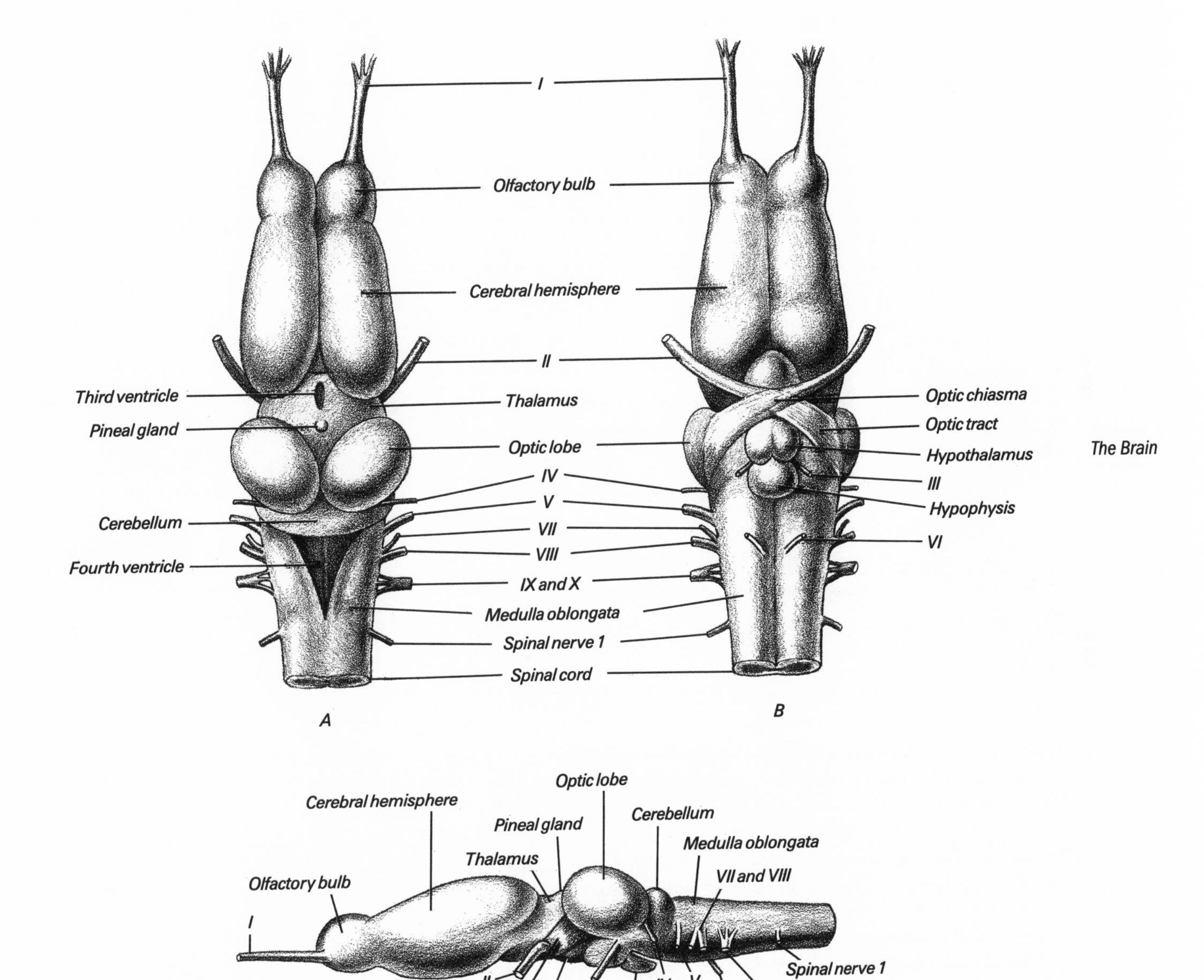

FIGURE 6-4
Dorsal (A), ventral (B), and lateral (C) views of the brain and the stumps of the cranial nerves.

A pair of large **optic lobes (optic tectum),** which make up the major part of the third brain region, the **mesencephalon,** are caudal to the diencephalon. The transverse band of nervous tissue caudal to the optic lobes is the **cerebellum.** This is the most conspicuous part of the **metencephalon,** the fourth region. The **medulla oblongata** follows the cerebellum and continues to the spinal cord. It constitutes the last of the five brain regions, the **myelencephalon.** It has a thin, vascular roof beneath which lies a large triangular cavity, the **fourth ventricle.** As you pick off the roof you may see another choroid plexus hanging from it.

2. Ventral and Lateral Aspects of the Brain

Cut across the medulla oblongata, and also across the cranial nerves, leaving as long stumps of the nerves as possible attached to the brain. As you carefully lift out the brain, you will see that a process from its ventral surface extends into a pocket on the floor of the skull. Chip away some additional bone and try to remove this process intact.

The Brain

Identify in ventral and lateral views the parts described above (Fig. 6-4, B and C). In addition, notice that the **optic nerves** cross as they reach the brain to form an **optic chiasma** from which **optic tracts** lead to optic lobes. The floor of the diencephalon caudal to the optic chiasma is the **hypothalamus,** and the **hypophysis,** or **pituitary gland,** attaches to it. If you do not see the gland, it may have become detached and may still be in the pocket in the floor of the skull. This gland produces a variety of hormones that affect growth, electrolyte and water balance, color changes, the activity of the thyroid and adrenal glands, and that control the production of hormones by the ovary and testis.

3. The Ventricular System of the Brain

Nick the walls of the cerebral hemispheres and you will see the first two, or lateral, **ventricles.** Each connects by a small **interventricular foramen** with the third ventricle, which you have seen in the diencephalon. The third ventricle connects by a **cerebral aqueduct** with the fourth ventricle, which lies in the metencephalon and myelencephalon. Unnumbered ventricles lie in the optic lobes. In the living animal, all of them are filled with a lymphlike **cerebrospinal fluid,** which is secreted by small vascular networks known as **choroid plexuses.**

Two of these are located in the thin roofs of the diencephalon and myelencephalon. This fluid helps to nourish the brain, and some escapes to circulate between the central nervous system and its meninges. By giving the brain some buoyancy in it, the cerebrospinal fluid also helps to support the brain and to buffer it against mechanical injury.

4. Brain Functions

Brain functions are very complex and are not fully understood, but meticulous tracing of neuron pathways in microscopic sections, and experiments involving the destruction of parts, the stimulation of parts, and the recording of electrical activity have given us many clues to brain activity. The cerebral hemispheres of the frog are primarily olfactory centers, which receive and integrate impulses from the nose and send out motor impulses to lower parts of the brain. There are no direct neuronal pathways from the cerebrum to the spinal cord as there are in mammals. Frogs whose cerebral hemispheres have been destroyed are more sluggish than normal, but, if appropriately stimulated, they will feed, jump, swim, and generally behave like normal frogs.

A variety of sensory fibers end in the thalamus, but few sensory impulses are relayed from it to the cerebral hemispheres. Presumably some integration occurs there.

The hypothalamus, like that in all vertebrates, is a very important center for the control of the autonomic nervous system and hence of visceral activity, including blood pressure, water and salt balance, gut movements, and much of the activity of the hypophysis.

The optic lobes clearly constitute that part of the brain governing most behavioral responses. They receive not only most of the fibers in the optic tract, but also neuronal projections from most of the other sense organs. Integration of different sensory modalities occurs in the optic lobes, and motor impulses initiated in them are sent to many parts of the brain and spinal cord. Electrical stimulation of certain parts of the lobes causes specific behavioral responses. In general, the optic lobes appear to have many of the functions performed by the cerebrum of mammals.

The cerebellum is a center for motor coordination, as it is in all vertebrates. It is much smaller in frogs than in most fishes, or in higher vertebrates, probably because locomotor activity is limited to little more than walking and kicking motions of the hind leg. There is little twisting or turning of the body, and there are few movements in many planes.

The medulla oblongata is another center for the control of visceral activity, including breathing and swallowing movements and the rate of heart beat.

5. Behavior

Frogs, like other animals, have complex behavioral patterns needed for their survival, such as flight from predators, feeding, and breeding. These have been studied intensively in recent years. Flight and feeding behavior, which can be demonstrated in the laboratory, will illustrate the complexities of the integrative functions of the central nervous system. Sight is the predominant sense in flight and feeding of terrestrial frogs and toads; but olfactory, lateral line, and cutaneous clues are more important in *Xenopus*. Try moving your hand toward a frog or toad, and then try dangling a small piece of meat or an insect on a thread in front of the animal. Observe the difference in behavior.

Whereas the image of a large moving object on the retina illicits avoidance reactions, a small moving pattern initiates prey-catching behavior. The frog turns toward the prey, moves slowly to snapping distance, and flicks out its tongue and draws the prey into its mouth. Mechanical stimulation by objects in the mouth initiates swallowing. Finally, the frog scrapes the snout clean with the forefeet.

This sequence of events involves many sense organs and parts of the central nervous system. The first distinction between relevant and irrelevant stimuli occurs in the retina of the eye (Exercise 7). Retinal images are projected point for point to the contralateral optic lobe. Some optic-lobe neurons respond primarily to large images, and especially to ones moving toward the animal. These neurons initiate escape reactions. Other neurons respond primarily to small images and initiate motor responses that cause the frog to turn toward the prey and move toward it. Other groups of neurons in the optic lobes and subtectal areas ventral to the lobes respond primarily to tactile stimuli, some to vibratory ones, and others to several other types of stimuli. Certain of these become active as the prey is caught and send motor impulses to appropriate motor centers in the brain, including the medulla oblongata, where swallowing movements are initiated. By placing stimulating electrodes in specific parts of the optic lobes, investigators have elicited all of the phases of feeding: turning, snapping, swallowing, and face cleaning. Other studies have demonstrated additional complexities. A center in the thalamus has an inhibitory influence on prey-catching behavior, and one in the cerebrum releases the inhibition of the thalamic center.

E. THE CRANIAL NERVES

The Cranial Nerves

The two largest cranial nerves, the olfactory and the optic, you have already seen. With the aid of Fig. 6-4, identify the stumps of as many of the others as possible. There are ten pairs in frogs, but some are small and may not be found. The ninth and tenth separate from each other after they leave the skull. A summary of the distribution of the nerves follows.

I. Olfactory nerve This consists of the axonlike processes of the olfactory cells (the receptors for smell), which are in the lining of the nasal cavity (Exercise 7). In the frog, they are collected into a grossly visible bundle because the olfactory bulb is some distance caudal to the nose. In most vertebrates, because the bulb is adjacent to the nose, the processes are short and not usually seen.

II. Optic nerve This is a sensory nerve that carries impulses from the retina of the eye.

III. Oculomotor nerve This is a motor nerve. Most of its fibers go to some of the extrinsic muscles that move the eyeball (inferior oblique, inferior rectus, superior rectus, and medial rectus; Exercise 7). One branch of the oculomotor nerve carries parasympathetic fibers to the ciliary body within the eye. A few sensory fibers— those that provide a feedback mechanism on the degree of contraction of the muscles—return to the brain in the oculomotor nerve. A few of these fibers return to the brain in most of the other motor nerves also.

IV. Trochlear nerve This nerve supplies another extrinsic ocular muscle (superior oblique).

V. Trigeminal nerve This is a mixed nerve that carries motor fibers to the jaw muscles, and returns sensory fibers from the skin covering the head and from part of the mucous membrane lining the mouth.

VI. Abducens nerve This nerve supplies the remaining extrinsic ocular muscles (lateral rectus, retractor bulbi).

VII. Facial nerve This too is a mixed nerve. Its motor fibers go to muscles controlling the hyoid apparatus, and most of its sensory fibers return from parts of the tongue and from some of the lining of the buccopharyngeal cavity. Many of these sensory fibers come from taste buds.

VIII. Statoacoustic nerve This nerve is composed of sensory fibers returning from the parts of the ear concerned with both equilibrium and hearing to the brain.

IX. Glossopharyngeal nerve This is a mixed nerve. Most of its motor fibers supply certain small throat muscles, and most of its sensory fibers return from part of the tongue and buccopharyngeal lining. Many of the sensory fibers, like those of the facial nerve, come from taste buds. A few parasympathetic fibers in the glossopharyngeal and facial nerves supply tear glands.

X. Vagus nerve The vagus supplies motor fibers to muscles that control the laryngotracheal chamber, and parasympathetic fibers to many of the visceral organs. Many sensory fibers from these regions return in the vagus. One branch of the vagus, which supplies the cucullaris muscle of the shoulder (Exercise 2), is comparable to the eleventh cranial nerve, the accessory, of higher vertebrates. Frogs lack a distinct twelfth cranial nerve, or hypoglossal nerve, which supplies tongue muscles in higher vertebrates. Comparable fibers are included in the first spinal nerve of frogs.

Sense Organs

RECEPTIVE CELLS are cells that are activated by a very slight change in a particular environmental modality to which they are attuned, and they in turn initiate an impulse in nerve fibers with which they connect. There are many types: some are sensitive to various kinds of mechanical deformation (touch, pressure, sound waves), some to light, some to temperature changes, and some to changes in the chemical environment. Most receptive cells are widely scattered throughout the body, but some are aggregated, along with other tissues, into conspicuous **sense organs**—the nose, eye, and ear. The associated tissues of sense organs support and protect the receptor cells and often amplify the environmental stimuli, or direct them toward the receptor cells.

A. THE NOSE

The **external nostrils,** or **nares,** are a pair of small openings near the front of the head that contain valves that open and close during the breathing cycle (Exercise 3). Each one leads into a **nasal cavity,** which can be found by skinning one side of the top of the head from the front of the jaw caudally to the eye, and then removing much of the overlying nasal bone (Fig. 7-1). The nasal cavity is rather large and somewhat triangular. An **internal nostril,** or **choana,** leads from its caudolateral corner into the roof of the mouth, and a conspicuous bulge in its floor, known as the **olfactory eminence,** partly divides the cavity into a groove-shaped medial chamber beside the mid-dorsal line and a broader lateral chamber.

Each nasal cavity has a dual function; it is part of the passageway for air movement to and from the lungs, and its lining contains many **olfactory cells** receptive to chemical changes in the air. These cells are of an unusual type known as neurosensory cells because they combine the properties of both receptive cells and neurons. Long cytoplasmic processes extending from them form the **olfactory nerve,** which leads directly to the brain (Fig. 7-1). The buccopharyngeal cavity also is part of the olfactory mechanism because its rhythmic expansion and contraction constitutes the pump that circulates air across the olfactory epithelium.

A small, inconspicuous group of special olfactory cells, which constitute the **vomeronasal organ of Jacobson,** lie in the most cranial part of the medial chamber in terrestrial frogs. The vomeronasal organs of vertebrates primarily detect odor bearing particles, which have been deposited in the environment by other members of the same species. These particles, called **pheromones,** are important in intraspecific social interactions.

B. THE EYE

The eyes of fishes are protected, cleansed, and moistened by the surrounding water, but in terrestrial vertebrates, eyelids and tear glands perform these functions. **Tear glands** in the frog are inconspicuous, yet prominent **upper** and **lower eyelids** are well developed in ranid frogs. Eyelids are very small in the aquatic *Xenopus.* The lower lid is a transparent membrane, which resembles the nictitating membrane of many terrestrial vertebrates but is probably not homologous to it. Carefully cut away the eyelids. A layer of epithelium, the **conjunctiva,** extends from the

The Eye

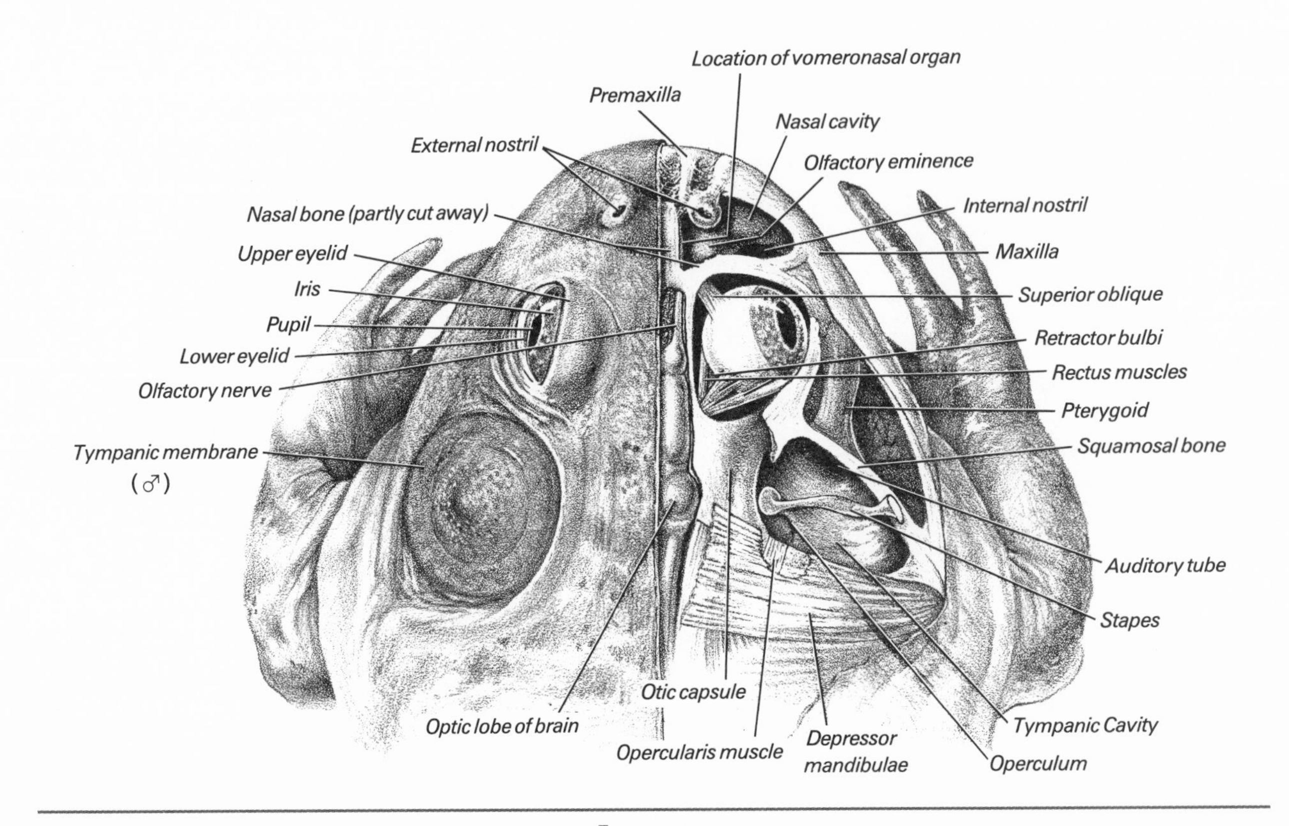

FIGURE 7-1
Dorsal view of the sense organs of *Rana.*
The nasal cavity, the orbit, the tympanic cavity, and the cranial cavity have been exposed on the right side.

underside of the lids over the surface of the transparent cornea. Next enlarge the orbit by cutting away part of the frontoparietal bone medial to the top of the eye, and then carefully pick away connective tissue and vessels from around the eyeball. As you do this, you will see the **extrinsic ocular muscles** that move the eyeball (Fig. 7-1). The individual muscles are difficult to discern in such a small animal, but if you dissect carefully, you may be able to distinguish two **oblique muscles** extending from the cranial part of the orbit to the top and underside of the eyeball; four **rectus muscles** extending from the caudal part of the orbit to the top, underside, medial and caudal surfaces of the eyeball; a **retractor bulbi** muscle lying beneath the rectus muscles. Contraction of the retractor bulbi pulls the eyeball deeper into the orbit and may even cause it to bulge into the roof of the mouth. The eyelids move across the eye as it is retracted. Lift up the eyeball and notice the sheet of muscle forming the floor of the orbit. This is the **levator bulbi,** and its contractions pushes the eyeball up.

The Eye

Major features of the eyeball are shown in Fig. 7-2. If you are dissecting a large frog you will be able to see many of these by removing, with a pair of fine scissors, a tangential slice from the dorsal surface of the eyeball. Do not try to cut through the lens. If the eyeball is left in the orbit during the dissection there will be less chance of crushing and injuring delicate structures, but the head should be submerged in water. The wall of the eyeball consists of three layers of tissue. The outer layer, known as the **fibrous tunic,** is a dense connective tissue that forms the supporting wall of the eyeball. The portion of the fibrous tunic on approximately the medial two-thirds of the eyeball is opaque and is known as the **sclera,** and that part of the fibrous tunic on the lateral one-third constitutes the transparent **cornea** through which light enters the eye. Some cartilage is present in the sclera. Beneath the fibrous tunic lies the **vascular tunic:** a pigmented and very vascular membrane most of which forms the **choroid** located between the sclera and the third layer, the **retina.** The choroid nourishes the retina, and its pigment absorbs excess light, thereby preventing a blurring of the image. The vascular tunic continues in front of the lens as the **iris.** Muscles within the iris regulate the size of the **pupil,** and hence the amount of light entering the eye. The part of the vascular tunic around the periphery of the lens, together with a nonnervous and inconspicuous portion of the retina, form the **ciliary body.** Minute fibers extend from the ciliary body to the lens, which they help to hold in place. The retina appears whitish in preserved specimens, and often it has collapsed and pulled away from the choroid. It contains on the surface next to the choroid the photoreceptive **rods** and **cones.** In life a semigelatinous **vitreous body** fills the large chamber between the retina and the lens, and a watery **aqueous humor** fills the **anterior** and **posterior chambers** on each side of the iris in front of the lens.

The frog's eye is well adapted to the animal's requirements and mode of life. The position of the eyes on

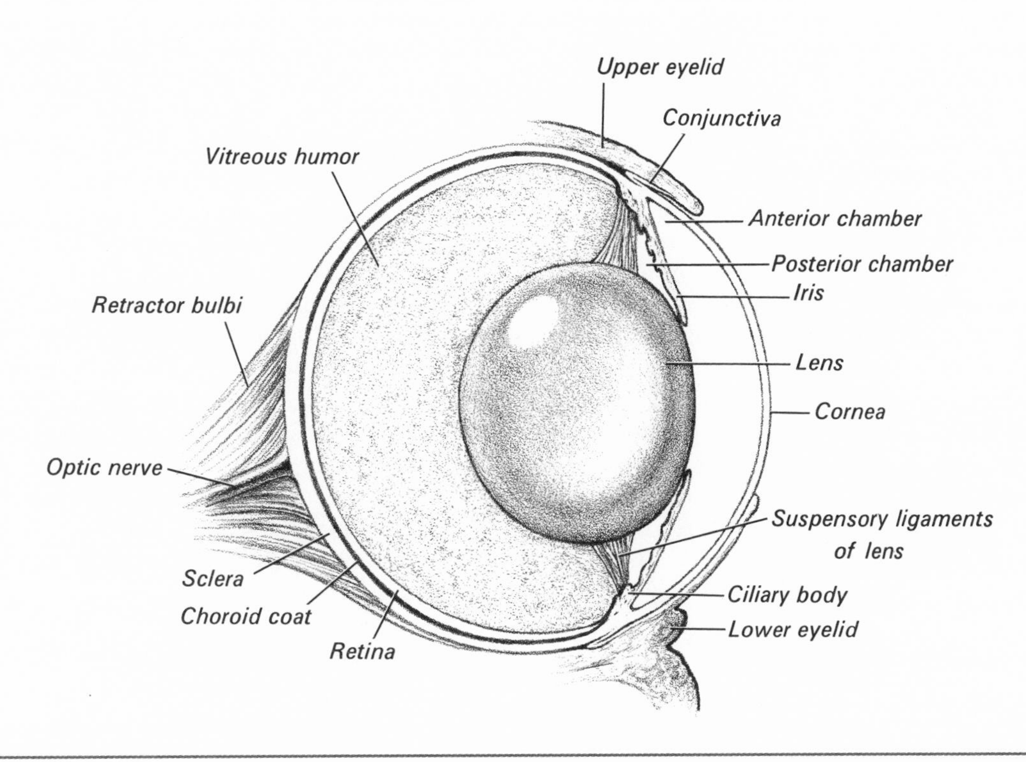

FIGURE 7-2
Diagrammatic vertical section of the eye (enlarged about six times).

top of the head, and the relatively large cornea and large lens of each, give frogs a very wide-angled visual field. Light rays entering the eye, are bent, or refracted, to form a sharp image upon the retina. In fishes, larval amphibians, and *Xenopus,* little refraction occurs in the cornea because its index of refraction is similar to that of water. Most refraction takes place as light passes through the lens, whose spherical shape gives it maximum refractive powers. At metamorphosis, and in adult terrestrial amphibians, the lens is flattened somewhat on its corneal surface. Its refractive powers are reduced because some refraction now occurs as light rays pass from the air through the denser cornea. Mammals have an even flatter lens. When the frog's eye is at rest, the eye is focused on moderately distant objects. **Accommodation** for a closer object is accomplished by inconspicuous **protractor lentis muscles,** associated with the ciliary body, that pull the lens away from the retina. This mechanism resembles the focusing mechanism of a camera more than does the accommodation mechanism of a mammal, in which the thickness of the lens is changed. Frogs cannot focus sharply on very distant objects, but they need not because they do not actively hunt prey; rather, they wait for organisms to come within their range of vision.

Embryonically the retina of the eye develops as an outgrowth of the diencephalon of the brain. It is not surprising, therefore, that it is a complex structure containing not only the receptive cells but also several layers of neurons that are interconnected in complex ways. Considerable processing of the image occurs at this level. A number of receptive cells converge, through intermediate cells, upon **ganglion cells** whose processes extend through the optic nerve to the brain. The ganglion cells have a certain base rate at which they send impulses to the brain. When an image stimulates the receptive cells, the associated ganglion cell is excited and increases its rate of discharge, and more impulses go to the brain. Simultaneously, because of the complex interconnections within the retina, adjacent areas are inhibited, and ganglion cells in them decrease their rate of discharge. These mechanisms accentuate borders of images and facilitate detecting the movement and velocity of an image across the retina. Different ganglion cells also have different field sizes; that is, the area of the retina that will stimulate them. These size differences provide information that enables the animal to distinguish size and shape. As a whole, the frog's eye is well adapted to detect those qualities that are particularly important for its survival.

The Ear

C. THE EAR

The ear of most frogs is adapted to receive airborne sound waves of biological importance to the animals, such as the mating calls of males and distress or warning calls. *Xenopus,* an aquatic species, presumably is adapted to hear underwater sounds by means of bone conduction, but its auditory apparatus has not been studied carefully. In *Rana,* a large **tympanic membrane** lies on the surface of the head caudal to each eye. It constitutes the **external ear,** and it responds to airborne vibratory waves by slight back-and-forth movements. It is distinctly larger in male bullfrogs than in females (Fig. 1-3), but there is no sexual difference in its size in leopard frogs. *Xenopus* lacks a tympanic membrane, but if the skin in the tympanic area is carefully removed you will find a thin plate of cartilage beneath it that probably receives pressure waves in the water.

Remove skin from the head above and behind the tympanic membrane. Cut around the periphery of the membrane, or the comparable plate of cartilage in *Xenopus,* and look into the **middle ear,** or **tympanic cavity.** A slender rod of cartilage and bone, the **stapes,** or **columella,** extends from the tympanic membrane (or surface cartilage in *Xenopus*) medially across the caudodorsal part of the cavity to the **otic capsule** (Fig. 7-1). The otic capsule is the part of the skull that contains the inner ear (Exercise 1). To see the area more clearly, remove the tympanic membrane and the ring of cartilage to which it is attached, and also remove muscles dorsal and caudal to the region. These muscles are the temporalis, depressor mandibulae, and cucullaris (Exercise 2, Fig. 2-1). A band of muscle known as the **opercularis** may be noticed extending from the deep surface of the suprascapula and scapula cranially and ventrally to the otic capsule. Save it. As you expose more of the tympanic cavity, notice that it connects with the roof of the buccopharyngeal cavity by way of the **auditory,** or **eustachian, tube.** Probe to verify this. Air pressure must be the same on each side of the tympanic membrane if it is to respond to sound waves; the auditory tube permits this equalization. Next carefully cut away some of the skull bones lying dorsal to the median part of the tympanic cavity and trace the stapes and opercularis muscles to the otic capsule. The medial end of the stapes expands slightly and fits into an **oval window,** or **fenestra vestibuli,** located on the lateral side of the otic capsule. The stapes transmits vibrations from the tympanic membrane to the inner ear. Pressure amplification, which is necessary to set up vibrations in the liquids of the inner ear, is achieved by the difference in size between the tympanic membrane and oval window, because nearly all of the energy that impinges on the membrane is concentrated at the much smaller oval window.

The opercularis muscle of ranid frogs attaches to the lateral surface of the otic capsule just caudal to the oval window. Detach it and carefully clear the lateral surface of the capsule. Notice that the part of the wall of the otic capsule to which the opercularis muscle is attached forms a distinct, roundish bone known as the **operculum.** The functional significance of the opercular apparatus of adult frogs is not certain. Recent studies support an earlier hypothesis that the opercular apparatus detects low frequency seismic vibrations from the ground. These would alert the animal to potential dangers, including the approach of some predators. The tympanic membrane and stapes function above 1000 Hz and detect communication signals from other frogs. It is of interest in this connection that terrestrial salamanders, which are mute, lack a tympanic membrane but have the opercular apparatus.

By chipping away some of the bone of the otic capsule you can see parts of the inner ear, but it is not easy to dissect it carefully with the instruments at your disposal. The chief part is the **membranous labyrinth** (Fig. 7-3), which is a complex of membranous canals and sacs filled with a lymphlike endolymph. Receptive cells are located in the swelling, or **ampulla,** at one end of each of the three **semicircular ducts,** and in the sac-shaped **utriculus** and **sacculus.** A large calcareous secretion, known as an **otolith,** overlies the receptors in the **utriculus** and **sacculus**. Collectively, these receptors initiate stimuli concerned with equilibrium, that is, with the position of the body in space, changes in position, and motion in various planes.

The membranous labyrinth is surrounded by a perilymphatic liquid, called perilymph, some of which transmits sound waves to receptors in the **papilla amphibiorum** and a **basilar papilla,** which is located in a part of the sacculus. It is probable that the papilla amphibiorum receives low frequency vibrations and the basilar papilla higher ones. Pressure waves are finally released by a sac of perilymph that protrudes into the cranial cavity. A unique feature of the frog ear is the extension of a part of the membranous labyrinth known as the **endolymphatic duct** into the cranial cavity and its caudal continuation beside the spinal cord. Portions of it containing calcareous granules have been seen as the **paravertebral lime sacs** surrounding the points of emergence of the ventral branches of the spinal nerves (Exercise 6).

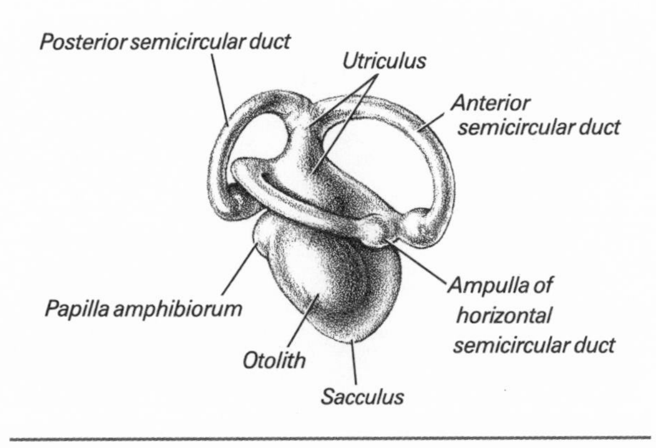

FIGURE 7-3
Lateral view of the right membranous labyrinth (enlarged about six times).

The Ear

Glossary of Vertebrate Anatomical Terms

THIS GLOSSARY IS A BASIC VOCABULARY of anatomical terms that students will encounter in using *Anatomy and Dissection of the Fetal Pig, Anatomy and Dissection of the Rat, Dissection of the Frog,* and many other exercises dealing with vertebrates. Most entries include the term, its pronunciation, its derivation, and its definition.

The pronunciation is given by a simple phonetic spelling in brackets. Stressed syllables are marked by a prime (′); others are separated by hyphens. Long vowels are indicated by the macron (¯); short ones, by the breve (˘).

The classical derivation in parentheses typically includes an abbreviation of the language of the word, the original word in italics, and the meaning of the original word. Usually only the nominative singular is given for Greek and Latin nouns, but the genitive (gen.) sometimes is used if it is closer to the root. Greek and Latin verbs usually are given in the first person present tense because this is closer to the root than the infinitive (but the meaning is given in the infinitive). The present or past participle (pres. p. or p. p.) is given when one of them is closer to the root. Many Greek and Latin words have a combining form that is used when the word is used in combination with other words. Combining forms are indicated by a hyphen before or after the word, for example, L. **Inter-** = between, as in interstitial.

When the English and classical terms are identical, the term is not repeated in the derivation, for example, **Acetabulum** (L. = vinegar cup). When two or more successive terms use the same root, the derivation of the root is given only for the first term. The origin of many repetitive terms is given only under the first entry. For example, **Ligamentum arteriosum** is defined the way this combination of terms is used, but the derivations of ligamentum and arteriosum will be found under **Ligament** and **Artery**. Names of individual muscles that are descriptive of an easily recognized feature of the muscle (its shape or attachments) are not given, but the component parts of less familiar ones are included. For example, the omohyoid muscle is not listed, but **Omo-** and **Hyoid** are. Similarly, names of blood vessels and nerves that simply state the organ supplied are omitted, but the organs are usually listed.

This glossary is not exhaustive. The pronunciation, derivation, and meaning of additional terms can be found in unabridged and medical dictionaries.

Abdomen [ab'dō-men] (L. from *abdo* = to conceal).
The part of the body containing the visceral organs, limited in mammals to the part caudal to the diaphragm.

Abducens nerve [ab-du'senz] (L. leading away, from *ab-* = away from + *duco*, pres. p. *ducens* = leading).
The sixth cranial nerve; carries motor fibers to an extrinsic ocular muscle.

Abduction [ab-dŭk'shŭn] (L. *duco*, p. p. *ductus* = to lead).
Muscle action that carries a body part away from a point of reference, often the midventral line of the body.

Accessory nerve.
The eleventh cranial nerve of amniotes. It carries motor fibers to certain branchiomeric shoulder muscles: the trapezius and sternocleidomastoid muscles.

Acetabulum [as-ĕ-tab'yū-lŭm] (L. = vinegar cup, from *acetum* = vinegar).
The cup-shaped socket in the pelvic girdle that receives the head of the femur.

Achilles tendon.
See **Tendon of Achilles.**

Acromion [ă-krō'mē-on] (Gr. *akron* = tip + *omos* = shoulder).
A process on the scapula with which the clavicle articulates in species with a well-developed clavicle.

Adduction [ă-dŭk'shŭn] (L. *ad-* = toward + *duco*, p.p. *ductus* = to lead).
Muscle action that pulls a body part toward a point of reference, often the midventral line of the body.

Adrenal gland [ă-drē'năl] (L. *ad-* = toward, beside + *ren* = kidney).
An endocrine gland located cranial to the kidney (mammals) or on its ventral surface (frogs). Its medullary hormone helps the sympathetic nervous system adjust the body to stress; its cortical hormones help regulate sexual development and the metabolism of minerals, carbohydrates, and proteins. It is also called the *suprarenal gland.*

Allantois [ă-lan'tō-is] (Gr. *allos* = sausage + *eidos* = appearance).
An extraembryonic membrane of reptiles, birds, and mammals. It develops as an outgrowth of the embryonic hindgut. It accumulates waste products. Its vascularized wall serves as a gas exchange organ in embryonic reptiles and birds, and it contributes to the formation of the placenta in eutherian mammals. Its base develops into the urinary bladder of adult mammals.

Alveolus [al-vē'ō-lŭs] (L. = small cavity).
One of a group of small, thin-walled, and vascularized sacs at the termination of the mammalian respiratory tree where gas exchange occurs.

Amnion [am'nē-on] (Gr. = the membrane around the fetus).
The innermost of the extraembryonic membranes. It envelops the fetus and contains amniotic fluid.

Amniote [am'nē-ōt].
A vertebrate whose embryo has an amnion; a reptile, bird, or mammal.

Amphibian [am-fi'bē-ăn] (Gr. *amphi* = both, double + *bios* = life).
A frog, salamander, or other member of the ancestral class of terrestrial vertebrates. Usually it is aquatic as a larva and terrestrial as an adult.

Ampullary gland [am-pul'lăry] (L. *ampulla* = small vessel).
A small gland associated with the terminal end of the ductus deferens in male rats. It contributes to the seminal fluid.

Anamniote [an-am'nē-ōt] (Gr. *an* = without + *amnion* = the membrane around the fetus).
A vertebrate without an amnion; a fish or amphibian.

Antebrachium [an-te-brā'kē-ŭm] (L. *ante* = before + *brachium* = upper arm).
The forearm.

Anterior.
A direction toward the front or belly surface of a human being; sometimes also used for the head end of a quadruped, but *cranial* is a more appropriate term.

Anterior chamber.
The space within the eyeball between the cornea and iris. It is filled with aqueous humor.

Anterior commissure.
An olfactory commissure within the cerebrum. It is located just rostral to the third ventricle.

Antrum [an'trŭm] (Gr. *antron* = a cave).
An enclosed cavity within an organ, such as the antrum in a secondary follicle in the mammalian ovary.

Anura [an-yūr'a] (Gr. *an* = without + *oura* = tail).
The amphibian order to which frogs and toads belong.

Anus [ā'nŭs] (L. = the seat, anus).
The caudal opening of the digestive tract in a mammal.

Aorta [ā-ōr'tă] (Gr. *aorte* = great artery).
The major artery carrying blood from the heart to the body. It is sometimes called the dorsal aorta to distinguish it from the ventral aorta of a fish, which carries blood from the heart to the gills.

Aortic valve.
A set of three semilunar-shaped folds at the base of the mammalian aorta. It prevents a backflow of blood into the left ventricle.

Aponeurosis [ap-ō-nū-rō'sis] (Gr. *apo* = from + *neuron* = sinew, nerve).
A sheetlike tendon of a muscle.

Appendix [ă-pen'diks] (L. *appendo* = to hang something on).
A dangling extension of an organ, such as the vermiform appendix at the end of the caecum of some mammals.

Aqueduct of Sylvius (*Franciscus Sylvius*, Dutch Anatomist, 1614–1672).
See **Cerebral aqueduct.**

Aqueous humor [ā'kwē-ŭs hyū'mer] (L. *aqua* = water + *humor* = fluid).
A watery liquid in the anterior and posterior chambers of the eyeball. It is secreted by the ciliary body.

Arachnoid [ă-rak'noyd] (Gr. *arachne* = spider + *eidos* = appearance).
One of three meninges of mammals. It is located between the dura mater and pia mater, and is connected to the pia mater by many strands, which give it a "spider web" appearance.

Arbor vitae [ar'bōr vīt'ē] (L. *arbor* = tree + *vita* = life).
The tree-shaped configuration of white matter within the mammalian cerebellum.

Archinephric duct [ar'ki-nef-rik] (Gr. *arche* = origin, beginning + *nephros* = kidney).
The duct, common to many fishes and amphibians, that drains the kidney. It also transports sperm in the male of many vertebrates. It becomes the ductus deferens in a male mammal. It is also called the *mesonephric duct.*

Areola [ă-rē'ō-lă] (L. = small space).
A small space or area, such as the small, round protuberances on the surface of a pig's chorion.

Arrector pili [ă-rek'tōr pi'li] (L. the raisor, from *arrectus* = upright + *pili* = hairs).
One of the small muscles in the skin of mammals that attach onto the hair follicles and raise the hairs.

Artery [ar'ter-ē] (L. *arteria* = artery).
A vessel that carries blood away from the heart. The blood is usually rich in oxygen, but it may be low in oxygen, as, for example, in the pulmonary arteries, which carry blood from the heart to the lungs.

Artiodactyla [ar'ti-ō-dak-til-a] (Gr. *artios* = even + *daktylos* = finger or toe).
The mammalian order that includes hoofed mammals with an even number of toes: pigs, deer, cattle.

Atlas [at'las] (Gr. mythology, a god who supported the Earth upon his shoulders).
The first cervical vertebra that supports the skull.

Atrioventricular valve [ā'trē-ō-ven-trik'yū-lar] (L. *atrium* = entrance hall + *ventriculus* = a small belly).
The valve between an atrium and ventricle of the heart. It prevents the backflow of blood from the ventricle into the atrium. The right atrioventricular valve of mammals has three cusps, or folds, and is also called the *tricuspid valve;* the left one has two cusps and is also called the *bicuspid,* or *mitral, valve.*

Atrium [ā'trē-um].
A chamber, such as the atrium of the heart, that receives blood from the sinus venosus or veins.

Auditory tube [aw'di-tōr-ē] (L. *audio*, p. p. *auditus* = to hear).
The tube that connects the tympanic (middle ear) cavity and the pharynx. It equalizes air pressure on each side of the tympanic membrane. Also called the *Eustachian tube.*

Auricle [aw'ri-kl] (L. *auricula* = little ear).
The external ear flap or pinna; also the ear-shaped lobe of a mammalian atrium.

Autonomic nervous system [aw-tō-nom'ik] (Gr. *autos* = self + *nomos* = rule, law).
An involuntary part of the nervous system (it "rules" itself.) It supplies motor fibers to glands and visceral organs.

Axillary [ak'sil-ār-ē] (L. *axilla* = armpit).
Pertaining to the armpit, e.g., the axillary artery.

Axis [ak'sis] (L. = axle, axis).
The second cervical vertebra of a mammal. Rotation of the head occurs between the atlas and axis.

Axon [ak'son] (Gr. = axle, axis).
The long, slender process of a neuron specialized to conduct nerve impulses a considerable distance.

Azygos vein [az'ĭ-gos] (Gr. *a* = without + *zygon* = yoke).
An unpaired vein in mammals. It drains most of the intercostal spaces on both sides of the body.

Basophil [bā'sō-fil] (Gr. *baso-* = alkaline + *phileo* = to love, to have an affinity for).
A leukocyte whose cytoplasmic granules take alkaline stains and appear blue; the rarest form of leukocytes.

Biceps [bī'seps] (L. *bi* = two + *ceps* = head).
A structure with two heads, such as the biceps brachii muscle, which is two-headed in human beings.

Bicornuate [bī-kōr'nū-āt] (L. *cornu* = horn).
A structure with two horns, such as a bicornuate uterus.

Bladder.
A membranous sac in which liquid may accumulate, e.g., the urinary bladder.

Blind spot.
See **Optic disk.**

Blood.
The liquid circulating in the blood vessels, consisting of a liquid plasma and cellular elements.

Bone.
The hard skeletal material of vertebrates consisting of an extracellular matrix of collagen fibers. It is mineralized with calcium phosphate crystals by bone forming cells.

Bowman's capsule. *(Sir William Bowman,* British Anatomist, 1816 - 1832.)
See **Glomerular capsule.**

Brachial [brā'kē-ăl] (L. *brachium* = upper arm).
Pertaining to the upper arm, e.g., brachial artery, coracobrachialis muscle.

Brachial plexus.
The network of nerves that supplies the shoulder and arm.

Brain.
The enlarged, cranial portion of the central nervous system. It is the major integrative center of the nervous system.

Braincase.
The cartilages and bones that encase the brain; also called the *cranium.*

Branchiomeric [brang'kē-o-mēr'ik] (Gr. *branchia* = gill + *meros* = part).
Pertaining to muscles or other structures associated with or derived from the visceral arches and gills.

Broad ligament.
Mesentery in female mammals. It anchors the reproductive tract to the dorsal body wall.

Bronchus [brong'kŭs] (Gr. *bronchos* = windpipe).
A branch of the trachea, which enters the lungs.

Buccal [bŭk'ăl] (L. *bucca* = cheek).
Pertaining to the mouth, e.g., buccal cavity.

Bulbourethral gland [bŭl'bō-yū-rē-thrăl] (L. *bulbus* = a bulbous root + Gr. *ourethra* = urethra).
A gland in male mammals near the base of the penis. It contributes to the seminal fluid. Also called *Cowper's gland.*

Caecum [sē'kŭm] (L. *caecus* = blind).
A pouch at the beginning of the large intestine. It is very long in many herbivores, including the rat, and houses bacteria and protozoa that digest cellulose.

Calcaneus [kal-kā'nē-ŭs] (L. = heel).
The large proximal tarsal bone that forms the "heel bone" in mammals.

Calyx, pl. calyces [kā'liks, kal'i-sēz] (Gr. *kaylx* = cup).
A cuplike compartment, such as one of the renal calyces, or subdivisions of the renal pelvis within the kidney.

Canaliculi [kan-ă-lik'yū-lī] (L. = small canals).
Minute canals in the bone matrix containing the slender processes of bone-forming cells.

Canine tooth [kā'nīn] (L. *canis* = dog).
The large pointed tooth in mammals. It is located caudal to the incisor teeth. In humans, the crown of the canine resembles an incisor, but its root is much larger. It is absent in the rat and enlarged in the pig.

Capillary [kap'i-lār-ē] (L. *capillus* = hair).
One of the minute, thin-walled blood vessels connecting small arteries to small veins. Exchanges between the blood and interstitial fluid occur across capillary walls.

Cardiac [kar'dē-ak] (Gr. *kardia* = heart).
Pertaining to the heart or its vicinity.

Carotid artery [ka-rot'id] (Gr. *karotides* = large neck artery, from *karoo* = to put to sleep, because compressing the artery leads to unconsciousness).
One of the arteries supplying blood to the head.

Carotid body.
A small enlargement at the junction of the external and internal carotid arteries. It is a chemoreceptor monitoring the levels of oxygen and carbon dioxide in the blood.

Carpal [kar'păl] (Gr. *karpos* = wrist).
One of the small bones forming the wrist.

Cartilage [kar'ti-lij] (L. *cartilago* = gristle).
A firm but elastic skeletal tissue whose matrix contains proteoglycan molecules that bind with water. Cartilage occurs in all embryos and in the skeleton of adult terrestrial vertebrates where firmness, smoothness, and flexibility are needed.

Caudal [kaw'dăl] (L. *cauda* = tail).
Pertaining to the tail or to a direction toward the tail in a quadruped.

Cecum.
See **Caecum.**

Cell body.
The part of a nerve cell, or neuron, that contains the nucleus. It does not include the axon and dendrites.

Central canal.
The cavity in the center of the spinal cord. It is continuous with the fourth ventricle in the brain and contains cerebrospinal fluid.

Central nervous system.
The part of the nervous system consisting of the brain and spinal cord.

Centrum.
See **Vertebral body.**

Cephalic [se-fal'ik] (Gr. *kephale* = head).
Pertaining to the head, e.g., brachiocephalic artery.

Cerebellum [ser-ĕ-bel'um] (L. = small brain).
The dorsal part of the metencephalon. It is an important center for muscular coordination and equilibrium.

Cerebral aqueduct [se-rē'brăl] (L. *cerebrum* = brain + *aqua* = water).
The narrow passage within the brain connecting the third and fourth ventricles; also known as the *aqueduct of Sylvius.*

Cerebral hemisphere.
One of two hemispheres that form most of the cerebrum. They are the major integration centers in the brain of mammals.

Cerebral peduncle [pe-dung'kl] (L. *pedunculus* = little foot).
One of a pair of neuronal tracts, which can be seen on the ventral surface of the mammalian brain. It carries impulses from the cerebrum to motor centers in the brain stem and spinal cord.

Cerebrospinal fluid.
A lymphlike fluid that circulates within and around the central nervous system and helps to protect and nourish it.

Cerebrum [se-rē'brŭm].
The major part of the telencephalon. It includes the cerebral hemispheres and a group of deep nuclei.

Cervical [ser'vĭ-kal] (L. *cervix*, gen. *cervicis* = neck).
Pertaining to the neck, e.g., cervical vertebrae.

Cervix [ser'viks].
The neck of an organ, e.g., the cervix of the uterus.

Choana [kō'an-ă] (Gr. *choane* = funnel).
One of the paired openings between the nasal cavity and pharynx; an internal nostril.

Choledochal duct.
See **Common bile duct.**

Chorda tympani [kōr'dă tim-pan'ē] (L. = cord + Gr. *tympanon* = drum).
A branch of the facial nerve crossing the malleus and traversing the tympanic cavity on its way to innervate certain taste buds on the tongue and salivary glands.

Chorion [ko'rē-on] (Gr. = skinlike membrane enclosing the fetus).
The outermost extraembryonic membrane of reptiles, birds, and mammals.

Choroid [kō'royd] (Gr. *chorioeides* = resembling a chorion).
The highly vascularized middle tunic of the eyeball lying between the retina and fibrous tunic.

Choroid plexus [plek'sŭs] (L. *plexus* = network).
Vascular tufts that project from a thin layer of brain tissue and protrude into a ventricle of the brain. They secrete the cerebrospinal fluid.

Chromatophore [krō-mat'ō-fōr] (Gr. *chromo* = color + *phoros* = bearing, from *pherein* = to bear).
A cell in nonmammalian vertebrates that contains pigment granules.

Ciliary body [sil'ē-ar-ē] (L. *cilium* = eyelash).
A part of the vascular tunic of the eyeball attached to the lens. Its muscle fibers regulate accommodation, and its secretory cells produce the aqueous humor.

Clavicle [klav'i-kl] (L. *clavicula* = small key).
The collar bone. It extends between the scapula and sternum in species in which it is well developed.

Cleido- [klī-do'] (Gr. *kleis*, gen. *kleidos* = key, clavicle).
A root referring to the clavicle; used in combination with other terms, e.g., cleidobrachialis muscle.

Clitoris [klit'ō-ris] (Gr. *kleitoris* = hill).
The small erectile organ of female mammals. It corresponds to much of the penis in a male mammal.

Cloaca [klō-ā'kă] (L. = sewer).
A chamber in nonmammalian vertebrates that receives the termination of the digestive, urinary, and genital tracts.

Coagulating gland.
A gland closely associated with the vesicular gland in male rats. It contributes to the seminal fluid.

Coccyx [kok'siks] (Gr. *kokkyx* = cuckoo).
The several fused caudal vertebrae of human beings. As a whole, the coccyx resembles the shape of the bill of a cuckoo. It does not reach the body surface but serves as the attachment site for some muscles.

Cochlea [kok'lĕ-a] (L. = snail shell).
Spiral part of the inner ear of mammals containing the auditory receptors and associated structures.

Coeliac artery [sē'lē-ak] (Gr. *koilia* = belly).
A branch of the aorta that supplies the cranial abdominal viscera, including the spleen, stomach, and liver.

Coelom [sē'lom] (Gr. *koiloma* = a hollow).
A body cavity that is completely lined by serosa, an epithelium of mesodermal origin.

Collagen [kol'lă-jen] (Gr. *kolla* = glue + *genos* = descent).
Minute protein fibers that form most of the extracellular material in connective tissues.

Colliculus [ko-lik'yū-lŭs] (L. = little hill).
One of four small elevations on the dorsal surface of the mammalian mesencephalon, which are centers for certain optic (superior colliculi) and auditory (inferior colliculi) reflexes.

Colon [kō′lon] (Gr. *kolon* = large intestine).
Most of the large intestine. Depending on the species, it extends from the small intestine or caecum to the cloaca or anus.

Columella [kol-yū-mel′a] (L. = small column).
A term frequently used for the rod-shaped stapes of nonmammalian terrestrial vertebrates. See also **Stapes.**

Commissure [kom′i-syūr] (L. *commissura* = a joining together).
A neuronal tract that interconnects structures on the left and right sides of the central nervous system.

Common bile duct.
The principal duct carrying bile to the intestine. It is formed by the confluence of hepatic ducts from the liver and, when present, the cystic duct from the gallbladder.

Concha [kon′kă] (Gr. *konkhe* = seashell).
One of several folds within the mammalian nasal cavity, which increase surface area; also called a turbinate bone.

Condyle [kon′dīl] (Gr. *kondylos* = knuckle).
A rounded articular surface, such as an occipital condyle.

Condyloid process.
The process of a mammalian mandible. It bears the facet for the jaw joint.

Conjunctiva [kon-jūnk-tī′va] (L. *conjunctus* = joined together).
The epithelial layer that covers the surface of, and fuses with, the cornea. It continues over the inner surface of the eyelids.

Connective tissue.
A widespread body tissue characterized by an extensive extracellular matrix containing fibers and cells. It includes collagenous and elastic connective tissue, fat, cartilage, and bone.

Conus arteriosus [kō′năs ar-ter-ē-ō′sas] (L. = cone).
A chamber of the heart in fishes and amphibians into which the single ventricle discharges blood. It contributes to the bases of the pulmonary trunk and aorta in mammals.

Coracoid [kōr′ă-koyd] (Gr. *korax*, gen. *korakos* = crow + *eidos* = appearance).
The bone forming the caudoventral part of the pectoral girdle in nonmammalian terrestrial vertebrates. It is reduced to a process (coracoid process) resembling a crow's beak in therian mammals.

Cornea [kōr′nē-ă] (L. *corneus* = horny).
The transparent part of the eyeball through which light passes. It is part of the fibrous tunic.

Coronary ligament [kōr′o-năr-ē] (L. *corona* = crown).
Peritoneum that bridges the gap between the liver and the diaphragm in mammals.

Coronary vessels.
Blood vessels that supply the heart musculature. During part of their course, they encircle the heart between the atria and ventricles.

Coronoid process.
The uppermost process of the mammalian mandible to which certain jaw muscles attach.

Corpus [kōr′pŭs] (L. = body).
The main body or part of an organ.

Corpus callosum [ka-lō′sum] (L. *callosus* = hard).
The large commissure interconnecting the two cerebral hemispheres in mammals.

Corpus cavernosum penis [kav-er-nō′sum pē′nis] (L. *caverna* = hollow space + *penis* = tail, penis).
One of a pair of columns of erectile tissue forming much of the penis.

Corpus luteum [lu-te′um] (L. *luteus* = yellow).
A yellowish endocrine gland within the ovary that develops from the ovulated follicle. The principal hormone it produces is progesterone.

Corpus spongiosum penis [spŭn-je′ō-sum] (Gr. *spongia* = sponge).
A column of erectile tissue that surrounds the penile portion of the urethra and forms the glans penis at the tip of the penis.

Cortex [kōr′teks] (L. = bark).
A layer of distinctive tissue on the surface of many organs, e.g., the cerebral cortex.

Costal [kos′tăl] (L. *costa* = rib).
Pertaining to the ribs, e.g., the costal cartilages.

Cowper's gland (*William Cowper*, British anatomist, 1666–1709).
See **Bulbourethral gland.**

Cranial [krā′nē-ăl] (Gr. *kranion* = skull).
Pertaining to the cranium; also a direction toward the head.

Cranium.
The skull, especially the part encasing the brain.

Cremasteric pouch [krē-mas-ter′ik] (Gr. *kremaster* = suspender).
Layers of the body wall that suspend the testis within the scrotum.

Crus, pl. **crura** [krūs, krū′ră] (L. = lower leg).
The lower leg or shin.

Cucullaris muscle [kyū′kū-lar-is] (L. *cucullus* = cap, hood).
A branchiomeric muscle in fishes and amphibians covering the craniodorsal part of the shoulder. It gives rise to the mammalian trapezius and sternocleidomastoid groups of muscles.

Cutaneous [kyū-tā′nē-ŭs] (L. *cutis* = skin).
Pertaining to the skin.

Cystic duct [sis′tik] (Gr. *kystis* = bladder).
The duct of the gallbladder. It joins the hepatic ducts to form the common bile duct.

Decussation [dē-kŭ-sā'shun] (L. *decusso*, p. p. *decussatus* = to divide crosswise in an X).

The crossing of neuronal tracts in the midline of the central nervous system.

Deferent duct.

See **Ductus deferens.**

Deltoid muscle [del'tōyd] (Gr. *deltoeides* = shaped like the letter delta, Δ).

A muscle crossing the lateral surface of the shoulder. It is shaped like the Greek letter *delta* in human beings.

Dendrite [den'drīte] (Gr. *dendron* = tree).

A highly branched and usually short process of a neuron that receives nerve impulses.

Dermis [der'mis] (Gr. *derma* = skin, leather).

The deeper layer of the skin. It is composed of dense connective tissue and develops from embryonic mesoderm.

Diaphragm [dī'ă-fram] (Gr. *dia* = through, across + *phragma* = partition, wall).

A mostly muscular partition between the thoracic and abdominal cavities in a mammal.

Diencephalon [dī-en-sef'ă-lon] (Gr. *enkephalos* = brain).

The brain region between the telencephalon and mesencephalon. It includes the epithalamus, thalamus, and hypothalamus.

Digit [dij'it] (L. *digitus* = finger).

A finger or toe.

Distal [dis'tăl] (L. *distalis* = situated away from the center).

The end of a structure most distant from its origin.

Dorsal [dōr'săl] (L. *dorsalis*, from *dorsum* = back).

A direction toward the surface of the back of a quadruped.

Duct [dŭkt] (L. *duco*, p. p. *ductus* = to lead).

A small, tubular passage carrying products away from an organ.

Ductus arteriosus [dŭk'tŭs ar-tēr'ē-ō-sŭs].

A connection in the fetal mammal between the pulmonary trunk and aorta, that permits much blood to bypass the embryonic lungs. It atrophies after birth and forms the ligamentum arteriosum.

Ductus deferens [def'er-enz] (L. *defero*, pres. p. *deferens* = to carry away).

The sperm duct of mammals and other amniotes; also called *vas deferens.*

Ductus venosus [vē-nō'sŭs].

A blood vessel in the liver of a fetal mammal. It permits much of the blood returning from the placenta in the umbilical veins to bypass the hepatic sinusoids and to enter the caudal vena cava directly.

Duodenum [dū-ō-dē'nŭm] (L. *intestinum duodenum digitorum*, *duodeni* = twelve fingers each).

The first part of the small intestine. In human beings, it is about 12 finger-breadths long.

Dura mater [dū-ră mā'ter] (L. = *durus* = hard + *mater* = mother).

The tough outer meninx surrounding the mammalian central nervous system.

Efferent ductules [ef'er-ent] (L. *ex* = out, away from + *fero*, pres. p. *ferens* = to carry).

Minute ducts in amphibians that carry sperm cells from the testis to the cranial tubules of the kidney; in amniotes, modified kidney tubules that lie in the head of the epididymis and transport sperm cells. These ducts are also called *vasa efferentia.*

Embryo [em'brē-ō] (Gr. *embryon* = embryo, from *en* = in + *bryo* = to swell).

An early stage in development of an individual. It is dependent for energy and nutrients from stored material within its egg or from the mother; i.e., embryos are not able to live on their own.

Endocrine gland [en'dō-krin] (Gr. *endon* = within + *krino* = to separate).

A ductless gland that discharges its secretion (a hormone) into the blood.

Eosinophil [ē-ō-sin'ō-fil] (Gr. *eos* = dawn + *philos* = fond).

A leukocyte whose cytoplasmic granules stain with eosin (an acid dye) and appear reddish.

Epaxial [ep-ak'sē-ăl] (Gr. *epi* = upon, above + *axon* = axle, axis).

Pertaining to those muscles and other organs that lie above or beside the dorsal half of the vertebral column.

Ependymal epithelium [ep-en'di-măl] (Gr. *ependyma* = garment).

The epithelial layer that lines the central canal of the spinal cord and the cavities in the brain.

Epidermis [ep-i-derm'is] (Gr. *epi* = upon + *derma* = skin).

The superficial layer of the skin. It is composed of a stratified, squamous epithelium whose outer layers are keratinized in terrestrial vertebrates. It is derived from the embryonic ectoderm.

Epididymis [ep-i-did'i-mis] (Gr. *didymoi* = testis).

A band-shaped group of tubules and a coiled duct, lying on the testis of mammals and in which sperm cells are stored. It evolved from a part of the kidney and kidney duct of ancestral amphibians and fishes.

Epiglottis [ep-i-glot'is] (Gr. *glottis* = entrance to the windpipe).

The flap of fibrocartilage that helps deflect food around the glottis of mammals during swallowing.

Epiphysis [e-pif'i-sis] (Gr. *physis* = growth).

(1) The end of a long bone in a young mammal.
(2) A thin outgrowth of the epithalamus in some fishes and amphibians, the distal end of which is sensitive to changes in light conditions. It corresponds to the mammalian pineal gland.

Epithalamus [ep'i-thal'ă-mŭs] (Gr. *thalamos* = inner chamber).
The roof of the diencephalon lying above the thalamus. Part of it is an olfactory center.

Epithelium [ep-i-thē'lē-um] (Gr. *thele* = delicate skin).
A tissue that is composed of tightly packed cells and covers body surfaces and lines cavities, including those of blood vessels and ducts.

Epitrichium [ep-i-trik'ē-ŭm] (Gr. *trichion* = small hair).
A layer of epithelium that lies upon the developing hairs in a mammal fetus.

Erectile tissue.
A tissue in an organ containing cavernous vascular spaces that may fill with blood and swell.

Esophagus [ē-sof'ă-gŭs] (Gr. *oisophagos* = gullet).
The part of the digestive tract between the pharynx and the stomach.

Eustachian tube (*Bartolommeo Eustachio,* Italian anatomist, 1520–1574).
See **Auditory tube.**

Exocrine gland [ek'sō-krin] (Gr. *ex* = out + *krino* = to separate).
A gland whose secretion is discharged through a duct onto a surface or into a cavity.

Extension [eks-ten'shŭn] (L. *extendo* = to stretch out).
A movement that carries a distal limb segment away from the next proximal segment, retracts a limb at the shoulder or hip, or moves the head or a part of the trunk toward the middorsal line.

External acoustic meatus [ă-kūs'tik mē-ā'tŭs] (Gr. *akoustikos* = pertaining to hearing + L. *meatus* = passage).
The external ear canal of amniotes extending from the body surface to the tympanic membrane.

External nostril.
See **Naris.**

Extrinsic [eks-trin'sik] (L. *extrinsecus* = from without).
A structure that originates from another structure or organ (e.g., extrinsic ocular muscles).

Extrinsic ocular muscles [ok'yū-lăr].
The group of small, strap-shaped muscles that extend from the wall of the orbit to the eyeball.

Facial [fā'shăl] (L. *facies* = face).
Pertaining to the face.

Facial nerve.
The seventh cranial nerve; innervates facial and other muscles associated with the second visceral arch and its derivatives, some salivary glands, and taste receptors on the front of the tongue.

Fallopian tube (*Gabriele Fallopio,* Italian anatomist, 1523–1562).
See **Uterine tube.**

Falx cerebri [falks se-rē'bri] (L. sickle + *cerebrum* = brain)
A sickle-shaped fold of dura mater in the longitudinal fissure between the two cerebral hemispheres.

Fascia [fash'ē-ă] (L. = band).
Sheets of connective tissue that lie beneath the skin (e.g., superficial fascia) or ensheathe groups of muscles (e.g., deep fascia) or organs.

Femur [fē'mŭr] (L. = thigh).
The thigh or bone within the thigh.

Fenestra [fe-nes'tră] (L. = window).
A windowlike opening in an organ.

Fenestra cochleae [kok'lĕ-ē] (L. *cochlea* = snail shell).
A small opening on the medial wall of the tympanic cavity of a mammal from which pressure waves, which have traveled through the inner ear, are released. It is often called the round window.

Fenestra vestibuli [ves-ti'būl-ē] (L. *vestibulum* = antechamber).
A small opening on the medial wall of the tympanic cavity into which the inner end of the stapes fits and propagates pressure waves into the inner ear. It is also called the *oval window.*

Fetus [fē'tŭs] (L. = offspring).
The unborn young of a mammal after it has nearly assumed the appearance it will have at birth (after the eighth week of gestation in human beings).

Fibroblast [fi'brō-blast] (L. *fibra* = fiber + Gr. *blastos* = bud).
An elongated and branching connective tissue cell that produces the intercellular matrix, including collagen fibers.

Fibrous tunic [tū'nik] (L. *tunica* = coat).
The tough connective tissue forming the outer wall of the eyeball; divided into the cornea and sclera.

Fibula [fib'yū-la] (L. = buckle, pin).
The slender bone on the lateral surface of the lower leg.

Fissure [fish'ūr] (L. *fissura* = cleft).
A deep groove or cleft in an organ, such as the brain or skull.

Flexion [flek'shŭn] (L. *flexio* = the bending).
A movement that brings a distal limb segment toward the next proximal one, advances a limb at the shoulder or hip, or bends the head or a part of the trunk toward the midventral line of the body.

Folia [fō'lē-ă] (L. = leaves).
Leaflike folds in an organ, such as those in the cerebellar hemispheres.

Foramen, pl. **foramina** [fō-rā'men, fō-ram'i-nă] (L. = opening).
An opening in an organ.

Foramen magnum [mag'nŭm] (L. *magnus* = large).
The opening in the base of the skull through which the spinal cord passes.

Foramen of Monro (*Alexander Monro, Sr.,* Scottish anatomist, 1697–1767).
See **Interventricular foramen.**

Foramen ovale [ō-văl'ē] (L. *ovalis* = oval).
A valved opening in the interatrial septum of a fetal mammal. It permits much of the blood (chiefly blood from the placenta and caudal part of the body) to pass from the right to the left atrium, thus, bypassing the lungs. It closes at birth and becomes the adult fossa ovalis.

Forebrain.
See **Prosencephalon.**

Fossa [fos'ă] (L. = ditch).
A shallow groove or depression in an organ.

Fossa ovalis [ō-vah'lis] (L. *ovalis* = oval).
An oval depression in the medial wall of the right atrium of an adult mammal. It is a vestige of the fetal foramen ovale.

Frontal [frŭn'tăl] (L. *frons,* gen. *frontis* = forehead).
(1) Pertaining to the forehead. (2) A plane of the body that passes through the frontal suture between the frontal and parietal bones; i.e., a median longitudinal plane passing from left to right.

Gallbladder [gawl'blad-er] (Old English *galla* = bile).
A small sac attached to the liver in most vertebrates. It stores and concentrates bile. The gallbladder is absent in rats.

Gametes [gam'ētz] (Gr. *gamete* = wife, *gametes* = husband).
The haploid germ cells, eggs and sperm.

Ganglion [gang'glē-on] (Gr. = small tumor, swelling).
A group of neuronal cell bodies that is a part of the peripheral nervous system in vertebrates.

Gastric [gas'trik] (Gr. *gaster* = stomach).
Pertaining to the stomach.

Gemelli muscles [jē-mel'lī] (L. = small twins).
Two small muscles situated deeply on the lateral surface of the hip in many mammals. They are fused with the obturatorius muscle in the pig.

Genio- [jē-ni'ō] (Gr. *geneion* = chin).
A combining term pertaining to the chin, as in the genioglossus muscle.

Girdles.
The skeletal elements in the body wall of vertebrates, that support the appendages.

Gland (L. *glans* = acorn).
A group of secretory cells. Exocrine glands discharge their secretion by a duct onto the body surface or into a cavity. Ductless endocrine glands discharge their secretion into the blood.

Glans clitoridis [glanz kli-tō'ri-dis] (Gr. *kleitoris* = hill).
The small mass of erectile tissue at the distal end of the clitoris in a female mammal.

Glans penis [pē'nis] (L. *penis* = tail, penis).
The bulbous, distal end of the penis in a mammal; part of the corpus spongiosum penis.

Glenoid fossa [glen'oyd] (Gr. *glene* = socket + *eidos* = resembling).
The socket that is located on the pectoral girdle of terrestrial vertebrates and receives the head of the humerus.

Glomerular capsule [glō-măr'yū-lăr] (L. *glomerulus* = little ball of yarn).
The thin-walled, expanded proximal end of a renal tubule. It surrounds a tuft of capillaries, the glomerulus. It is also called *Bowman's capsule.*

Glomerulus [glō-măr'yū-lŭs] (L. = little ball of yarn).
A dense network of capillaries that is surrounded by the glomerular capsule at the proximal end of a kidney tubule.

Glossopharyngeal nerve [glos'ō-fă-rin-jē-ăl] (Gr. *glossa* = tongue + *pharynx* = throat).
The ninth cranial nerve. It carries motor fibers to pharyngeal muscles derived from the third visceral arch, parasympathetic fibers to certain salivary glands, and returns sensory fibers from taste buds on part of the tongue and from the part of the pharynx near the base of the tongue.

Glottis [glot'is] (Gr. = opening of the windpipe).
A slitlike opening in the center of the larynx, including the space between the vocal cords.

Gluteal [glū'tē-ăl] (Gr. *gloutos* = buttock).
Pertaining to the buttocks, e.g., gluteal muscles.

Goblet cells.
Goblet-shaped, mucus-producing cells associated with the epithelial lining of the stomach, intestine, and upper respiratory tract.

Gonads [gō'nadz] (Gr. *gone* = seed).
A collective term for the testes and ovaries.

Graafian follicle [gră'fē-ăn] (*Reijnier de Graaf,* Dutch anatomist, 1641–1673).
See **Tertiary follicle.**

Gray matter.
Tissue in the central nervous system consisting primarily of cell bodies of neurons and unmyelinated nerve fibers.

Gubernaculum [gū'ber-nak'yū-lŭm] (L. = small rudder).
A cord of connective tissue that lies at the caudal end of the testis and guides the descent of the testis into the scrotum.

Gyrus [jī'rŭs] (Gr. *gyros* = circle).
One of the folds on the surface of a cerebral hemisphere.

Hair.
A filamentous epidermal structure in the skin of mammals. It consists primarily of keratinized cells. En masse, it helps to thermally insulate the body.

Hallux [hal'ŭks] (Gr. = big toe).
The first or most medial digit of the foot.

Harderian gland (*Johann Harder,* Swiss anatomist, 1656–1711).
A tear gland present in some mammals. It is located rostral and ventral to the eyeball. Also called *accessory lacrimal gland.*

Hard palate.
The part of the secondary palate that is supported by a horizontal shelf of bone in the mammalian skull and separates the oral from the nasal cavities.

Haversian system (*Clopton Havers,* British anatomist, 1650–1702).
See **Osteon.**

Heart.
The hollow muscular organ that pumps blood through the body.

Hemoglobin [hē-mō-glō'bin] (Gr. *haima* = blood + L. *globus* = globe).
Iron-containing, globular molecules that fill the red blood cells and bond reversibly with oxygen and some carbon dioxide.

Hepatic [he-pat'ik] (Gr. *hepar* = liver).
Pertaining to the liver.

Hepatic duct.
One of the ducts that drain the liver. Hepatic ducts join a cystic duct from the gallbladder to form the common bile duct in animals that possess a gallbladder.

Hepatic portal system.
A system of veins that drain the capillaries of the abdominal digestive organs and spleen and that lead to the sinusoids within the liver.

Hepatic vein.
One of the veins that drains the hepatic sinusoids and, in mammals, leads to the caudal vena cava.

Hindbrain.
See **Rhombencephalon.**

Hyaloid artery [hi'ă-loyd] (Gr. *hyolos* = glass + *eidos* = appearance).
The embryonic artery that passes through the vitreous body of the eyeball to supply the developing lens. It disappears before birth.

Hyobranchial apparatus [hī'ō-brang'kē-ăl] (Gr. *hyo* = a combining form referring to structures associated with the second visceral, or hyoid, arch + *branchia* = gill).
The complex of cartilages and bones that is derived from the hyoid and other visceral arches and which supports the tongue and the floor of the mouth and pharynx in nonmammalian terrestrial vertebrates. It is also called the hyoid apparatus in mammals.

Hyoid [hī'oyd] (Gr. *hyoeides* = resembling the letter *upsilon* = U-shaped).
(1) Pertaining to structures associated with the second visceral, or hyoid, arch. (2) The mammalian bone that is embedded in and supports the base of the tongue.

Hyoid apparatus.
See **Hyobranchial apparatus.**

Hypaxial [hī-pak'sē-ăl] (Gr. *hypo* = under + *axon* = axle or axis).
Pertaining to those muscles and other structures that lie in the body wall ventral to the dorsal half of the vertebral axis.

Hypobranchial [hī-pō-brang'kē-ăl] (Gr. *branchia* = gill).
Pertaining to muscles and other structures derived from structures that are located ventral to the gills in fishes.

Hypoglossal nerve [hī-pō-glos'ăl] (Gr. *glossa* = tongue).
The twelfth cranial nerve of amniotes. It carries motor fibers to the muscles of the tongue.

Hypophysis [hī-pof'i-sis] (Gr. *physis* = growth).
An endocrine gland that is attached to the ventral surface of the hypothalamus. It secretes many hormones that regulate a variety of physiological processes and other endocrine glands. Also called the pituitary gland.

Hypothalamus [hī-pō-thal'ă-mŭs] (Gr. *thalamus* = inner chamber).
The ventral part of the diencephalon. It lies ventral to the thalamus and is an important center for the integration of visceral functions.

Ileum [il'ē-ŭm] (L. = small intestine).
The caudal half of the small intestine of mammals. In human beings it is the part of the small intestine that lies against the ilium.

Iliac [il'ē-ak] (L. *ilium* = groin, flank).
Pertaining to a structure near the groin or flank, e.g., iliac artery.

Ilium [il'ē-ŭm] (L. = groin or flank).
The bone forming the dorsal part of the pelvic girdle of terrestrial vertebrates. It articulates with the sacrum.

Incisor tooth [in-sī'zŏr] (L. = the cutter, from *incido,* p. p. *incisum* = to cut into).
One of the front teeth of mammals lying rostral to the canine tooth. It is used for cutting and cropping and is very large in rodents.

Incus [ing'kŭs] (L. = anvil).
The anvil-shaped, middle auditory ossicle of mammals.

Inferior [in-fē'rē-or] (L. = lower).
A direction toward the feet of a human being.

Infundibulum [in-fŭn-dib'yū-lŭm] (L. = little funnel).
A funnel-shaped structure, such as the entrance into the oviduct or uterine tube.

Inguinal [ing'gwi-năl] (L. *inguen,* gen. *inguinis* = groin).
Pertaining to structures in or near the groin.

Inguinal canal.
A passage through the muscular and fascial layers of the body wall, leading from the abdominal cavity to the vaginal cavity of the scrotum. Ducts, blood vessels, and nerves to and from the testis travel through this passage.

Inner ear.
The portion of the ear that lies within the otic capsule of the skull and contains the receptive cells for equilibrium and hearing.

Insertion of a muscle.
The point of attachment of a muscle that moves the greater distance when the muscle contracts. It is usually the distal end of a limb muscle.

Integument [in-teg'yū-ment] (L. *integumentum* = covering).
The body covering consisting of the epidermis and the dermis. It often includes many accessory structures: glands, hair, feathers, scales. Also called skin.

Internal nostril.
See **Choana.**

Interneuron [in'ter-nū-ron] (L. *inter* = between + Gr. *neuron* = nerve, sinew).
A neuron that lies within the central nervous system and connects sensory and motor neurons or other interneurons. It is responsible for most of the integrative activity of the nervous system.

Interstitial cells [in-ter-stish'al] (L. *interstitium* = a small space between).
Groups of endocrine cells that lie between the seminiferous tubules of the testis and secrete the male sex hormone, testosterone.

Interstitial fluid.
A lymphlike fluid that lies in the minute spaces between the cells of the body.

Interthalamic adhesion [in-ter-tha-lam'ik] (Gr. *thalamos* = inner chamber).
The part of the mammalian thalamus that crosses the midline within the third ventricle. Also called the massa intermedia.

Interventricular foramen [in-ter-ven-trik'yū-lar fō-ra'men] (L. *ventriculus* = belly).
A foramen leading from one of the lateral ventricles of the brain to the third ventricle. Also called the *foramen of Monro.*

Intestine [in-test'tin] (L. *intestinus* = intestine).
The portion of the digestive tract between the stomach and anus or cloaca. It is the site of most digestion and absorption.

Intrinsic [in-trin'sik] (L. *intrinsecus* = on the inside).
A structure that is an inherent part of an organ, e.g., the ciliary muscles of the eyeball.

Iris [ī'ris] (Gr. = rainbow).
The pigmented part of the vascular tunic of the eyeball. It surrounds the pupil.

Ischiadic, ischiatic [is-ke-ad' (at) -ik] (Gr. *ischion* = hip).
Pertaining to a structure near the hip, e.g., the ischiadic nerve.

Ischium [is'kē-ŭm].
The bone forming the caudoventral part of the pelvic girdle.

Islets of Langerhans (*Paul Langerhans,* German physician, 1847–1888).
See **Pancreatic islets.**

Jacobson's organ (*Ludwig L. Jacobson,* Danish surgeon and anatomist, 1783–1843).
See **Vomeronasal organ.**

Jejunum [jě-jū'nŭm] (L. *jejunus* = empty).
Approximately the first half of the mammalian postduodenal small intestine. It is usually found to be empty in autopsies.

Jugular vein [jug'yū-lar] (L. *jugulum* = throat).
One of the major veins in the neck of mammals. It drains the head.

Keratin [ker'ă-tin] (Gr. *keras* = horn).
A horny protein synthesized by the epidermal cells of terrestrial vertebrates.

Kidney [kid'nē] (Middle English *kindenei* = kidney).
The organ that removes waste products, especially nitrogenous wastes, from the blood and produces urine.

Kidney tubule.
See **Renal tubule.**

Kneecap.
See **Patella.**

Labia [lā'bē-ă] (L. = lips).
Liplike structures, e.g., the genital labia of a female mammal.

Lacrimal [lak'ri-măl] (L. *lacrima* = tear).
Pertaining to structures involved with the production and transport of tears (e.g., lacrimal gland) or to structures located near these structures (e.g., lacrimal bone).

Lacuna [lă-kŭ'nă] (L. = a small hollow or space).
A small cavity, such as one of many in the bone matrix, in which a bone cell (osteocyte) lodges.

Lamella [lă-mel'ă] (L. = small plate, layer).
A thin plate or layer of tissue, such as the layers of matrix in bone.

Laryngotracheal chamber.
A chamber in the respiratory tract of amphibians into which the glottis leads and from which the lungs emerge. It is comparable to the larynx and trachea of mammals.

Larynx [lar′ingks] (Gr. = larynx).
The part of the respiratory tract of amniotes between the pharynx and trachea. It contains the vocal cords in mammals.

Lateral [lat′er-ăl] (L. *latus,* gen. *lateris* = side).
Pertaining to the side of the body.

Lens [lenz] (L. = lentil).
A refractive transparent body near the front of the eyeball. It is responsible for accommodation or focusing by changing its shape (mammals) or by moving toward or away from the retina (frogs).

Leukocyte [lū′kō′sīt] (Gr. *leukos* = clear, white + *kytos* = cell).
Any of several different types of white blood cells.

Lienic [lī′ĕ-nik] (L. *lien* = spleen).
Pertaining to the spleen, e.g., lienogastric artery.

Ligament [lig′ă-ment] (L. *ligamentum* = band, bandage).
(1) A band of dense connective tissue extending between certain structures, usually bones. (2) A mesentery extending between certain visceral organs.

Ligamentum arteriosum [lig′ă-men-tum ar-tēr′ē-ō-sum].
A band of dense connective tissue extending from the pulmonary trunk to the aorta in adult mammals. It is a remnant of the embryonic ductus arteriosus.

Linea alba [lin′ē-ă al′ba] (L. = white line).
A whitish strand of connective tissue in the midventral abdominal wall to which abdominal muscles attach.

Limbic system [lim′bik] (L. *limbus* = border).
A brain region that encircles the diencephalon and leads to the hypothalamus. It is important in regulating many behaviors related to species survival, including feeding and reproduction.

Lingual [ling′gwăl] (L. *lingua* = tongue).
Pertaining to the tongue.

Liver [liv′er] (Old English *lifer* = liver).
The large organ in the cranial part of the abdominal cavity. It secretes bile and plays a vital role in many metabolic processes, including processing many substances brought to it by the hepatic portal vein system, disassembling senescent red blood cells, detoxifying substances, and synthesizing many plasma proteins.

Lumbar [lŭm′bar] (L. *lumbus* = loin).
Pertaining to structures in the back between the thorax and pelvis, such as the lumbar vertebrae.

Lung [lŭng] (Old English *lungen* = lung).
One of the paired organs of terrestrial vertebrates in which gases are exchanged between the air and blood. It develops as an outgrowth from the floor of the pharynx.

Lymph [limf] (L. *lympha* = clear water).
A clear liquid that is derived from the interstitial fluid and flows in the lymphatic vessels. It lacks the red blood cells and most of the plasma proteins found in blood.

Lymph heart.
A pulsating part of lymphatic vessels of some amphibians and reptiles. It helps return lymph to the veins.

Lymph node.
Nodules of lymphatic tissue along the course of the lymphatic vessels of mammals. It is the site for the multiplication of many lymphocytes, the phagocytosis of many foreign particles, and the initiation of certain immune response.

Lymphocyte [lim′fō-sīt] (Gr. *kytos* = cell).
A leukocyte with a large, round nucleus and very little cytoplasm. Lymphocytes develop in lymph nodes, spleen, thymus, and other lymphoid tissues, and participate in immune responses.

Macrophage [mak′rō-fāj] (Gr. *makros* = large + *phagein* = to eat).
Large cell in the tissues. It phagocytoses foreign particles and participates in immune responses.

Malleus [mal′ē-ŭs] (L. = hammer).
The outermost of three auditory ossicles of a mammal. It transmits sound waves across the tympanic cavity; it is shaped like a hammer.

Mammal [mam′ăl] (L. *mamma* = breast).
A member of the vertebrate class characterized by hair and mammary glands.

Mammary gland [mam′ă-rē].
The milk-secreting, cutaneous gland that characterizes female mammals.

Mammary papilla [pă-pil′ă] (L. = nipple, pimple).
The nipple, or teat, of a mammary gland.

Mandibular cartilage [man-dib′yū-lăr] (L. *mandibula* = lower jaw).
A cartilaginous core of the bony lower jaw of many fishes, amphibians, and reptiles. It represents the ventral half of the first visceral arch of ancestral fishes.

Mandibular gland.
A mammalian salivary gland usually located deep to the caudoventral angle of the lower jaw.

Mandibular ramus [rā′mŭs] (L. = a branch).
The vertical part of the mandible caudal to the teeth.

Massa intermedia.
See **Interthalamic adhesion.**

Masseter muscle [mă-sē′ter] (Gr. *masseter* = chewer).
A large mammalian muscle that helps to close the jaws. It extends from the zygomatic arch to the mandible.

Mastoid process [mas′toyd] (Gr. *mastos* = breast + *eidos* = appearance).
A large, rounded process on the base of the mammalian skull located caudal to the external acoustic meatus; it is a point of attachment for certain neck muscles.

Meatus [mē-ā′tŭs] (L. = a passage).
A passage, such as the external acoustic meatus, that leads from the head surface to the tympanic membrane.

Meconium [mē-kō′nē-um] (Gr. *mekonion* = poppy juice).
Bile-stained debris in the fetal digestive tract. It is discharged shortly after birth.

Medial [mē′dē-al] (L. *medialis* = middle).
A direction toward the middle of the body.

Median [mē′dē-an] (L. *medianus* = middle).
Lying in the midline of the body.

Mediastinum [me′dē-as-tī-nŭm] (L. = middle septum).
The space in the mammalian thorax between the two pleural cavities. It contains the aorta, esophagus, pericardial cavity, heart, and the venae cavae.

Medulla [me-dūl′ă] (L. = core, marrow).
The central part, or core, of certain organs, as opposed to their surface region, or cortex.

Medulla oblongata [ob-long-gah′tă].
The myelencephalon, or caudal region of the brain, which is continuous with the spinal cord.

Melanin [mel′ă-nin] (Gr. *melas* = black).
A black pigment often contained in cells known as melanophores.

Meninx, pl. **Meninges** [mē′ningks, mĕ-nin′jēz] (Gr. = membrane).
A connective tissue membrane that surrounds the central nervous system.

Mesencephalon [mez-en-sef′ă-lon] (Gr. *mesos* = middle + *enkephalos* = brain).
The middle, or third, of five brain regions. It lies between the diencephalon and metencephalon and includes the colliculi (mammals) or optic lobes (frogs); also called the midbrain.

Mesentery [mez′en-ter-ē] (Gr. *enteron* = intestine).
(1) Any membrane-like double layer of serosa that extends from the body wall to visceral organs, or between visceral organs. (2) The particular membrane suspending the intestine.

Mesonephric duct.
See **Archinephric duct.**

Mesorchium [mez-ōr′kē-ŭm] (Gr. *orchis* = testis).
The mesentery suspending the testis.

Mesovarium [mez′ō-va′rē-um] (L. *ovarium* = ovary).
The mesentery suspending the ovary.

Metacarpal [met′ă-kar′păl] (Gr. *meta* = after + *karpos* = wrist).
One of the long bones in the palm of the hand located between the carpals and phalanges.

Metatarsal [met′ă-tar′săl] (Gr. *tarsos* = ankle).
One of the long bones of the foot located between the tarsals and phalanges.

Metencephalon [met′en-sef′ă-lon] (Gr. *enkephalos* = brain).
The fourth of the five brain regions. It lies between the mesencephalon and myelencephalon and includes the cerebellum and (in mammals) the pons.

Midbrain.
See **Mesencephalon.**

Middle ear.
The portion of the ear of terrestrial vertebrates that usually contains the tympanic cavity and one or three auditory ossicles. Airborne sound waves are transmitted from the body surface (usually from a tympanic membrane) across the middle ear cavity via the auditory ossicle (or ossicles) to the inner ear.

Middle ear cavity.
See **Tympanic cavity.**

Molar tooth [mō′lăr] (L. *molaris* = millstone).
One of the caudal teeth of mammals usually adapted for crushing or grinding food.

Monocyte [mon′ō-sīt] (Gr. *monos* = single + *kytos* = cell).
A leukocyte with a kidney-shaped nucleus. It is a precursor of a tissue macrophage.

Mouth [mowth] (Old English *muth* = mouth).
The cranial opening of the digestive tract. Also used for the oral cavity into which this opening leads.

Mucosa [myū-kō-să] (L. *mucosus* = slimy, mucous).
The lining of the digestive tract and many other hollow visceral organs. It consists of an epithelium, connective tissue, and sometimes a thin layer of smooth muscle.

Mucus [myū′kŭs] (L. = slime).
A slimy material secreted by some epithelial cells. It is rich in the glycoprotein mucin.

Muscle [mŭs-ĕl] (L. *musculus* = muscle, a little mouse).
(1) A contractile tissue responsible for most of the movements of the body or its parts. (2) A discrete group of muscle fibers with a common origin and insertion.

Myelencephalon [mī′el-en-sef′ă-lon] (Gr. *myelos* = core, spinal cord + enkephalos = brain).
The most caudal of the five regions of the brain. It consists of the medulla oblongata.

Mylohyoid [mī′lō-hī′ōyd] (Gr. *myle* = a mill + *hyoeides* = U-shaped).
A nearly transverse sheet of muscle extending between the two mandibles and the hyoid bone.

Myo- [mī´ō] (Gr. *mys*, gen. *myos* = muscle).
A prefix meaning muscle or musclelike.

Myoepithelium [mī´ō-ep-thē-lē-um].
Specialized epithelial cells containing contractile elements. Some surround the secretory cells of sweat glands and help to discharge sweat.

Myofibrils [mī-ō-fī´brilz] (L. *fibrilla* = a minute fiber).
Minute, longitudinal, contractile fibrils within a muscle fiber. They are barely visible with a light microscope and are composed of myofilaments.

Myofilaments [mī´ō-fil´ă-mentz].
Ultramicroscopic filaments of actin and myosin whose interactions are the basis of muscle contraction.

Naris, pl. **nares** [nā´ris, nā´res] (L. = nostril).
An opening from the outside into the nasal cavity; an external nostril.

Nasal [nā´zăl] (L. *nasus* = nose).
Pertaining to the nose, e.g., nasal bone, nasal cavity.

Nasal conchae [kon´kē] (L. = shells).
Folds within the nasal cavity of mammals. They increase the surface area.

Nasal meatus (L. = a passage).
An air passage in the mammalian nasal cavity between the nasal conchae, or between the conchae and nasal septum.

Nephron [nef´ron] (Gr. *nephros* = kidney).
The structural and functional unit of the kidney. It consists of a glomerulus and a renal tubule.

Nerve [nerv] (L. *nervus* = nerve, sinew).
A bundle of neuronal processes (axons) and their investing connective tissues. It extends between the central nervous system and peripheral organs, and is part of the peripheral nervous system.

Neural arch.
See **Vertebral arch.**

Neuron [nūr´on] (Gr. *neuron* = nerve, sinew).
A nerve cell. It is the structural and functional unit of the nervous system. It consists of dendrites, a cell body, and an axon.

Neuron tract.
A bundle of neuronal processes within the spinal cord or brain.

Neutrophile [nū´trō-fil] (L. *neutro* = neither + Gr. *philos* = affinity for).
A leukocyte with small cytoplasmic granules that stain only lightly with basic and acidic dyes. Neutrophiles are phagocytic and are the most abundant leukocytes.

Nictitating membrane [nik´ti-tāt-ing] (L. *nicto* = to beckon, wink).
A membrane in many terrestrial vertebrates that can slide across the surface of the eyeball. It is reduced to a vestigial semilunar fold in human beings.

Nipple [nip´l] (Old English *neb* = beak, nose).
A papilla that bears the openings of the ducts from the mammary gland.

Nostril [nos´tril] (Old English *nosus* = nose + *thyrl* = hole).
The opening into the nasal cavity from the body surface (external nostril, naris) or from the nasal cavity into the pharynx (internal nostril, choana).

Obturator foramen [ob´tū-rā-tŏr] (L. *obturo*, p. p. *obturatus* = to close by stopping up).
A large foramen in the mammalian pelvic girdle. The obturator muscles arise from the periphery of the foramen and close it.

Occipital [ok-sip´i-tăl] (L. *occiput* = back of the head).
Pertaining to the back of the head or skull.

Occipital condyle [kon´dīl] (Gr. *kondylos* = condyle, knuckle).
One of a pair of enlargements on the occipital bone on each side of the foramen magnum of amphibians and mammals, or a single enlargement ventral to the foramen magnum in most other vertebrates. They articulate with the cranial articular surface of the atlas.

Ocular [ok´yū-lăr] (L. *oculus* = eye).
Pertaining to the eye, e.g., the extrinsic ocular muscles that move the eyeball.

Oculomotor nerve [ok´yū-lō-mō-tŏr].
The third cranial nerve. It carries motor nerve fibers to most of the extrinsic muscles of the eyeball and autonomic nerve fibers into the eyeball.

Olecranon [ō-lek´ră-non] (Gr. = the tip of the elbow).
The proximal end of the ulna. It extends behind the elbow joint in mammals.

Olfactory [ol-fak´tŏ-rē] (L. *olfacio*, p. p. *olfactus* = to smell).
Pertaining to the parts of the nose involved in smelling.

Olfactory nerve.
The first cranial nerve, which consists of neurons returning from the nose to the olfactory bulb of the brain.

Omentum [ō-men´tŭm] (L. = fatty membrane).
The peritoneal membrane, often containing a great deal of fat. It extends between the body wall and stomach (i.e., greater omentum) or between the stomach and the liver and duodenum (i.e., lesser omentum).

Omo- [ō´mō] (Gr. *omos* = shoulder).
A combining form pertaining to the shoulder, e.g., omotransversarius muscle.

Oocyte [ō´ō-sīt] (Gr. *oon* = egg + *kytos* = cell).
An early stage in the development of the egg. The first meiotic division of the primary oocyte produces the secondary oocyte and a polar body.

Oogonium [ō-ō-gō'nē-ŭm] (Gr. *gone* = generation).
A very early stage in the development of an egg. It enlarges to become the primary oocyte.

Ootid [ō-ō-tid] (Gr. *ootidium* = a small egg).
The nearly mature egg, or ovum, after the second meiotic division has been initiated. In mammals, fertilization initiates the completion of this division.

Optic [op'tik] (Gr. *optikos* = pertaining to the eye).
Pertaining to the parts of the eye involved in vision.

Optic chiasma [kī-az'ma] (Gr. = cross, from the Greek letter *chi* = X).
The complete or partial decussation of the optic nerve fibers on the ventral surface of the diencephalon.

Optic disk.
A disk-shaped area on the retina to which the optic nerve attaches. It lacks photoreceptors; also called the "blind spot."

Optic lobe.
One of a pair of lobes on the dorsal surface of the mesencephalon in nonmammalian vertebrates. It is a major integrating center in these vertebrates.

Optic nerve.
The second cranial nerve. It carries impulses from the retina to the brain.

Optic tract.
The neuronal tract leading from the optic chiasma to the thalamus, or optic lobes, or both.

Oral cavity [ōr'ăl] (L. *os*, gen. *oris* = mouth).
The mouth cavity; also called the buccal cavity.

Orbit [or'bit] (L. *orbis*, gen. *orbitis* = circle).
A circular cavity on the side of the skull. It lodges the eyeball.

Origin of a muscle.
The point of attachment of a muscle that tends to remain in a fixed position when the muscle contracts; the proximal end of a limb muscle.

Ossicle [os'i-kl] (L. *ossiculum* = small bone).
Any small bone, such as one of the auditory ossicles.

Osteocyte [os'tē-ō-sīt] (Gr. *osteon* = bone + *kytos* = cell).
A mature bone cell surrounded by the matrix it has produced.

Osteon [os'tē-on].
A cylindrical, microscopic unit of bone consisting of concentric layers of bone matrix surrounding a central canal that contains blood and lymph vessels. Also called a *Haversian system.*

Ostium [os'tē-ŭm] (L. = entrance, mouth).
The entrance to an organ, such as the ostium tubae of the uterine tube.

Otic capsule [ō'tik] (Gr. *otikos* = pertaining to the ear).
The part of the skull surrounding the inner ear.

Otolith [ō'tō-lith] (Gr. *oto-* = ear + *lithos* = stone).
Calcareous granules within sacs of the inner ear that stimulate sensory cells over which they move when an animal moves or changes position.

Oval window.
See **Fenestra vestibuli.**

Ovarian follicle [ō-var'ē-an] (L. *ovarium* = ovary + *folliculus* = little bag).
A group of epithelial and connective tissue cells in the ovary. It surrounds and nourishes the developing egg. It is also an endocrine gland whose primary hormone is estrogen.

Ovary [ō'vă-rē].
One of a pair of female reproductive organs containing the ovarian follicles and eggs.

Oviduct [ō'vi-dŭkt] (L. *ovum* = egg + *ducere*, p. p. *ductus* = to lead).
The passage in females of nonmammalian vertebrates. It transports eggs from the coelom to the cloaca.

Oviparous [ō-vip'ă-rŭs] (L. *pario* = to bear).
To bear eggs. A pattern of reproduction found in frogs and many nonmammalian vertebrates that lay eggs. The embryos develop outside the body of the mother.

Ovisac [ō'vi-sak].
An enlargement at the caudal end of the oviduct in which eggs accumulate before they are released.

Ovulation [ov'yū-lā-shŭn].
The release of mature egg cells from the ovarian follicles and ovary into the coelomic cavity (in the frog), ovarian bursa (in the rat and pig), or infundibulum (in the human female and many other mammals).

Ovum [ō'vum].
The mature egg cell.

Palate [pal'ăt] (L. *palatum* = palate).
The roof of the mouth. See also **Secondary palate.**

Pampiniform plexus [pam-pin'i-fōrm] (L. *pampinus* = tendril + *forma* = shape).
A network of veins in mammals entwining the testicular artery as it approaches the testis.

Pancreas [pan'krē-as] (Gr. *pan* = all + *kreas* = flesh).
A large glandular outgrowth of the duodenum. It secretes many digestive enzymes. It also contains the endocrine pancreatic islets of Langerhans.

Pancreatic islets.
Small clusters of endocrine cells in the pancreas. They produce hormones that regulate sugar metabolism; also called the *islets of Langerhans.*

Papilla [pă-pil'ă] (L. = nipple).
A small conical protuberance.

Papilla amphibiorum.
The part of the amphibian inner ear receptive to low-frequency sound waves.

Paradidymis [par′ă-did′i-mis] (Gr. *para* = beside + didymoi = testis).
A small group of vestigial kidney tubules occurring in male mammals and located beside the epididymis.

Paraflocculus [par-ă-flok′yū-lŭs] (L. *flocculus* = a small tuft of wool).
A small lobe on the lateral surface of the cerebellum. It is conspicuous in the rat.

Parasympathetic [par-ă-sim-pa-thet′ik] (Gr. *syn* = with + *pathos* = feeling).
Pertaining to the parasympathetic part of the autonomic nervous system. In mammals, preganglionic parasympathetic neurons leave the central nervous system through certain cranial nerves and sacral spinal nerves. The parasympathetic system stimulates metabolic processes that absorb and store energy.

Parathyroid gland [par-ă-thī′royd] (Gr. *thyreos* = oblong shield + *eidos* = shape).
One of several endocrine glands of a terrestrial vertebrate. It is located dorsal to, near, or entwined with the thyroid gland. Its hormone, called parathormone, helps to regulate calcium and phosphate metabolism.

Parietal [pă-rī′ĕ-tăl] (L. *paries*, gen. *parietis* = wall).
Pertaining to the wall of some structure, e.g., parietal bone, parietal peritoneum.

Parotid gland [pă-rot′id] (Gr. *para* = beside + *otikos* = pertaining to the ear).
A large mammalian salivary gland located ventral to the external ear.

Patella [pa-tel′ă] (L. = small plate).
The mammalian kneecap, composed of bone. It is a sesamoid bone within the patellar tendon of the quadriceps muscle.

Pectoral [pek′tŏ-răl] (L. *pectoralis* = pertaining to the chest).
Pertaining to the chest, e.g., pectoral muscles, pectoral appendage.

Pelvic [pel′vik] (L. *pelvis* = basin).
Pertaining to basin-shaped structures, such as the pelvic girdle, or renal pelvis.

Penis [pē′nis] (L. = tail, penis).
The male copulatory organ of most amniotes.

Pericardial cavity [per-i-kar′dē-ăl] (Gr. *peri* = around + *kardia* = heart).
The portion of the coelom that surrounds the heart.

Pericardium.
The serosa that covers the surface of the heart (i.e., visceral pericardium) and forms part of the pericardial wall (i.e., parietal pericardium).

Periosteum [per-ē-os′tē-ŭm] (Gr. *osteon* = bone).
The vascularized and innervated connective tissue covering a living bone.

Peritoneal cavity [per′i-tō-nē-al] (Gr. *peritonaion* = to stretch over).
The portion of the mammalian coelom that houses the abdominal viscera.

Peritoneum.
The serosa that covers the abdominal visceral organs (i. e., visceral peritoneum) and lines the peritoneal cavity (i.e., parietal peritoneum).

Peroneus muscle [per-ō-nē′ŭs] (Gr. *perone* = pin, fibula).
One or more muscles located on the lateral side of the shin over the fibula and extending into the foot.

Phalanges [fă-lan′jēz] (Gr. = battle lines of soldiers).
Bones of the digits that extend beyond the palm of the hand or sole of the foot.

Pharynx [far′ingks] (Gr. = throat).
That part of the digestive tract that lies between the oral cavity and the esophagus; the crossing place of the digestive and respiratory tracts. Embryonically, lungs develop as outgrowths from the pharyngeal floor, and pharyngeal pouches develop from the lateral walls of the pharynx.

Phrenic [fren′ik] (Gr. *phren* = diaphragm).
Pertaining to the diaphragm, e.g., phrenic nerve, phrenic artery.

Pia mater [pī′ă mā′ter] (L. = tender mother).
The delicate, vascularized layer of connective tissue that invests the brain and spinal cord. It is the innermost of the three mammalian meninges.

Pineal gland [pin′ē-ăl] (L. *pineus* = pertaining to pine, from *pinus* = pine tree).
An endocrine gland that produces melatonin, especially under dark conditions. The functions of melatonin are not fully understood, but it has been implicated in adjusting physiological processes to diurnal and seasonal cycles.

Piriformis muscle [pir′i-fōrm-is] (L. *pirum* = pear + *forma* = shape).
A small, triangular muscle located on the medial side of the proximal end of the thigh. It is pear-shaped in human beings.

Pituitary gland.
See **Hypophysis.**

Placenta [plă-sen′tă] (L. = flat cake).
The apposition or union of parts of the uterine lining and fetal extraembryonic membranes through which food, respiratory gases, and waste products are exchanged between the mother and the fetus.

Plantaris muscle [plan′tār-is] (L. = pertaining to the sole of the foot).
A large muscle on the caudal surface of the crus of nonmammalian vertebrates. Its tendon runs onto the sole of the foot and extends the foot.

Platelet [plāt′let] (English, *small plate*).
An irregularly shaped fragment of the cytoplasm of a megakaryocyte in the blood. It is involved in blood clotting.

Platysma muscle [plă-tiz′mă] (Gr. *platys* = flat, broad + *-ma* = suffix indicating result of an action, e.g., of making something flat and broad).
Thin sheet of muscle underlying the skin of the neck in mammals.

Pleura [plūr′ă] (Gr. = side, rib).
The serosa that covers the lungs (i.e., visceral pleura) and lines the pleural cavities (i.e., parietal pleura).

Pleural cavities.
The paired coelomic spaces that enclose the lungs in mammals.

Pleuroperitoneal cavity [plūr-ō-per-i-tō-nē′al].
The combined potential peritoneal and pleural cavities in anamniotes. It contains the abdominal viscera and, if present, the lungs.

Plexus [plek′sŭs] (L. = network, braid).
A network of blood vessels or nerves, e.g., choroid plexus, brachial plexus.

Polar body [pō′lăr] (L. *polaris* = pole).
A small body located near the animal pole of a developing egg cell. It results from the unequal division of the egg during the first and second meiotic division.

Pollex [pol′eks] (Gr. = thumb).
The thumb.

Pons [ponz] (L. = bridge).
The ventral part of the mammalian metencephalon. Its most conspicuous surface feature is a bridgelike tract of transverse neuron fibers.

Popliteal [pop-lit′ē-ăl] (L. *poples*, gen. *poplitis* = knee joint).
The depression behind the mammalian knee joint.

Portal vein [pōr′tăl] (L. *porta* = gate, door).
A vein that carries blood from the capillaries of one organ to the capillaries of another organ rather than to the heart, e.g., the hepatic portal vein.

Posterior [pos-tēr′ē-ōr] (L. *poster* = after, following + *ior* = suffix indicating the comparative form of an adjective, e.g., more behind).
A direction toward the back of a human being. It is sometimes used for the tail end of a quadruped, but *caudal* is a more appropriate term.

Posterior chamber.
The space within the eyeball between the iris and lens. It is filled with aqueous humor.

Postganglionic motor neuron [post′gang-glē-on′ik] (Gr. *ganglion* = small tumor, swelling).
A neuron of the autonomic nervous system whose cell body lies in a ganglion in the peripheral nervous system.

Preganglionic motor neuron [prē′gang-glē-on′ik] (L. *pre-* = in front of).
A neuron of the autonomic nervous system whose cell body lies within the central nervous system and whose axon synapses with a postganglionic neuron.

Premolar tooth [prē-mō′lăr] (L. *molaris* = millstone).
One of several mammalian teeth that lie in front of the molar teeth and behind the canine tooth. They usually are adapted for a combination of cutting and grinding.

Prepuce [prē′pūs] (L. *praeputium* = foreskin).
The foreskin of a male mammal. It covers the glans penis.

Prosencephalon [prōs-en-sef′ă-lon] (Gr. *pro-* = before + *enkephalos* = brain).
The embryonic forebrain. It gives rise to the telencephalon and diencephalon.

Prostate [pros′tāt] (Gr. *prostates* = one who stands before).
An accessory genital gland of male mammals. It surrounds the urethra just before the urinary bladder and secretes much of the seminal fluid.

Protraction [prō-trak′shŭn] (L. *traho*, p. p. *tractus* = to pull).
Muscle action that moves the entire appendage of a quadruped forward.

Proximal [prok′si-măl] (L. *proximus* = nearest).
The end of a structure nearest its origin, e.g., the end of a limb that is closest to the girdle.

Pterygoideus muscle [ter′i-goyd-e-us] (Gr. *pteryx*, *pteryg-* = wing + *eidos* = resembling).
A jaw-closing muscle that arises from the pterygoid bone (frog) or pterygoid process on the underside of the skull (mammals) and inserts on the medial side of the lower jaw.

Pubis [pyū′bis] (L. = genital hair).
The cranioventral bone of the pelvic girdle of terrestrial vertebrates.

Pudendal [pyū-den′dăl] (L. *pudendum* = external genitals, from *pudendo* = to feel ashamed).
Pertaining to the region of the external genitals, e.g., pudendal artery.

Pulmonary [pŭl′mō-nār-ē] (L. *pulmo* = lung).
Pertaining to the lungs.

Pulmonary trunk.
The mammalian arterial trunk that leaves the right ventricle and soon divides into the two pulmonary arteries going to the lungs.

Pulmonary valve.
A set of three semilunar folds in the base of the pulmonary trunk. It prevents the backflow of blood into the right ventricle.

Pupil [pyū'pĭl] (L. *pupilla* = pupil).
The central opening in the iris through which light enters the eyeball.

Pylorus [pī-lōr'ŭs] (Gr. *pyloros* = gatekeeper).
The caudal end of the stomach, which contains a sphincter muscle.

Pyramidal system [pi-ram'i-dal] (Gr. *pyramis* = pyramid).
The direct neuronal motor pathway in mammals, from pyramid-shaped cell bodies in the cerebrum to motor neurons in the brain and spinal cord.

Quadrate cartilage [kwah'drāt] (L. *quadratus* = square).
A cartilage at the caudal end of the upper jaw of a frog with which the mandibular cartilage of the lower jaw articulates. It represents the dorsal half of the first visceral arch of ancestral fishes.

Radius [rā'dē-ŭs] (L. = ray, spoke).
The bone of the forearm that rotates around the ulna in most terrestrial vertebrates. It lies on the lateral side of the forearm when the palm is supine.

Ramus [rā'mŭs] (L. = branch).
A branch of a nerve or a blood vessel, also part of the mandible.

Rectum [rek'tŭm] (L. *rectus* = straight).
The caudal end of the large intestine of a mammal.

Renal [rē'năl] (L. *ren* = kidney).
Pertaining to the kidney, e.g., renal artery.

Renal portal system.
A system of veins that drains the capillaries of the hind legs and tail of most nonmammalian vertebrates and leads to capillaries around the kidney tubules.

Renal tubule.
A kidney tubule. It consists of a glomerular capsule and a tubule; the tubular part of a nephron.

Rete testis [rē'tē tes'tis] (L. = net + *testis* = testicle).
A network of small passages in the mammalian testis between the seminiferous tubules and the epididymis. They form a visible cord in the rat.

Retina [ret'i-nă].
The innermost layer of the eyeball. It contains pigment, receptive rods and cones, and associated neurons.

Retraction [rē-trak'shŭn] (L. *re* = backward + *tractio* = a pull).
Muscle action that pulls the entire appendage of a quadruped caudad.

Rhinal [rī'năl] (Gr. *rhis*, gen. *rhinos* = nose).
Pertaining to the nose.

Rhinal sulcus.
The furrow that separates the portion of the cerebrum dealing with olfaction from other areas of the brain.

Rhinencephalon [rī'nen-sef'ă-lon] (Gr. *enkephalos* = brain).
The primary olfactory portion of the cerebrum.

Rhombencephalon [rom-ben-sef'ă-lon] (Gr. *rhombus* = an oblique-angled equilateral parallelogram + *enkephalos* = brain).
The embryonic hindbrain. It develops into the metencephalon and myelencephalon.

Rodentia [rō-den'shē-ă] (L. *rodo*, pres. p. *rodens* = to gnaw).
The mammalian order, including the rat and other gnawing animals, that is characterized by two enlarged incisor teeth in the upper and lower jaws.

Rostral [ros'trăl] (L. *rostrum* = beak).
A direction toward the front of the head. This term is used for structures within the head.

Round ligament.
A round cord; specifically, a connective tissue cord that crosses the broad ligament in female mammals and connects the ovary with the body wall at the groin. It corresponds to the male gubernaculum.

Round window.
See **Fenestra cochleae.**

Sacculus [sak'yū-lŭs] (L. small sac).
The most ventral chamber of the inner ear. It contains an otolith and functions as a sensor of equilibrium.

Sacral vertebrae, sacrum [sa'krŭm] (L. = sacred).
The vertebrae, or fusion of two or more vertebrae and their ribs, with which the pelvis articulates.

Sagittal [saj'i-tăl] (L. *sagitta* = arrow).
A plane of the body that passes through the sagittal suture (between the parietal bones of the skull), e.g., a median, longitudinal plane passing from dorsal to ventral.

Salivary gland [sal'i-vār-ē] (L. *saliva* = saliva).
One of several glands that secrete saliva. Major ones in mammals are the parotid, mandibular, and sublingual glands.

Sarcolemma [sar'kō-lem-ă] (Gr. *sarx*, gen,. *sarkos* = flesh + *lemma* = husk).
The cell membrane of a muscle cell.

Sartorius muscle [sar-tōr'ē-ŭs] (L. *sartor* = tailor).
A narrow, diagonal muscle on the medial surface of the thigh. It is especially well developed in people

who sit cross-legged on the floor, such as tailors in preindustrial times, because it is used to rise.

Scapula [skap'yū-lă] (L. = shoulder blade).
The shoulder blade, or the element of the pectoral girdle that extends dorsally on the back.

Sclera [sklēr'a] (Gr. *skleros* = hard).
The opaque (white) portion of the fibrous tunic of the eyeball. Together with the cornea, it forms the outer wall of the eyeball.

Scrotum [skrō'tŭm] (L. = pouch).
The cutaneous sac that encases the paired mammalian testes when the testes are descended.

Sebaceous gland [sē-bā'shŭs] (L. *sebum* = tallow).
A gland in mammalian skin. It produces an oily or waxy secretion, which is usually discharged into a hair follicle.

Secondary palate.
The palate of mammals that separates the food and air passages. It consists of the hard palate, which separates the oral from the nasal cavities, and the fleshy soft palate, which separates the oral pharynx from the nasal pharynx.

Semicircular duct [sem'ē-sir-kyū-lăr] (L. *semi* = prefix denoting one-half).
One of three semicircular ducts in the inner ear. Each lies at a right angle to the others and detects changes in angular acceleration generated by turns of the head.

Seminal fluid [sem'i-năl] (L. *semen* = seed).
The fluid secreted by male reproductive ducts and accessory genital glands.

Seminal vesicle.
See **Vesicular gland.**

Seminiferous tubules [sem-i-nif'er-ŭs] (L. *semen,* gen. *seminis* = seed + *fero* = to carry).
Tubules within the testis in which sperm cells are produced.

Septum [sep'tum] (L. = partition).
A partition.

Septum pellucidum [pe-lū'si-dum] (L. *pellucidus* = clear, transparent).
A thin, vertical septum of nervous tissue in the brain ventral to the corpus callosum. It forms the medial wall of each lateral ventricle.

Serosa [se-rō'să] (L. *serosus* = watery).
The coelomic epithelium and underlying connective tissue that line body cavities and cover visceral organs, e.g., peritoneum, pleura, pericardium, and tunica vaginalis.

Serratus muscle [ser-āt'us] (L. = toothed like a saw).
A muscle with a serrated or saw-toothed border.

Sertoli cell (*Enrico Sertoli,* Italian histologist, 1842–1910).
Large epithelial cell in the seminiferous tubules. It plays a role in the maturation of the sperm cells.

Sesamoid bone [ses'ă-moyd] (Gr. *sesamon* = sesame seed + *eidos* = resembling).
A bone that develops in the tendon of some muscles near their insertion and facilitates the movement of the tendon across a joint, or changes the direction of the force of a muscle, e.g., the patella and the pisiform.

Sinus [sī-nŭs] (L. = a cavity).
A cavity or space within an organ.

Sinus venosus [vē-nō'sus] (L. *venosus* = pertaining to a vein).
The most caudal chamber of the heart of anamniotes and some reptiles. In frogs, it receives blood from the body and leads to the right atrium.

Skin.
See **Integument.**

Skull [skŭl] (Old English *skulle* = a bowl).
The group of bones and cartilages that encase the brain and major sense organs and form the upper jaw and face. The lower jaw sometimes is considered to be a part separate from the skull.

Soft palate.
The fleshy partition in mammals between the oral pharynx and nasal pharynx.

Somatic [sō-mat'ik] (Gr. *somatikos* = bodily).
Pertaining to the body wall and appendages, as opposed to the internal (visceral) organs, e.g., somatic muscles, somatic skeleton.

Sperm cell [sperm] (Gr. *sperma* = seed, sperm cell).
One of the mature male gametes; also called spermatozoa.

Spermatid [sper'mă-tid] (Gr. *idion,* a diminutive ending).
A cell resulting from the second meiotic division of a secondary spermatocyte. It develops into a sperm cell.

Spermatocyte [sper'mă-tō-sīt] (Gr. *kytos* = cell).
The primary spermatocyte results from the growth and division of a spermatogonium, and the secondary spermatocyte results from the first meiotic division of the primary spermatocyte.

Spermatogonium [sper'mă-tō-gō-nē-ŭm] (Gr. *gone* = generation).
The stem cell that multiplies mitotically and gives rise to sperm cell-forming cells.

Spermatozoa.
See **Sperm cell.**

Spinal [spī'năl] (L. *spina* = spine, thorn).
Pertaining to a spine-shaped structure, often the spine or vertebral column, e.g., spinal cord, spinal nerve.

Spinous process.
The dorsal, spinelike process of the vertebral arch.

Splanchnic [splangk'nik] (Gr. *splanchnon* = gut, viscus).
Pertaining to structures that supply the gut or visceral organs, such as the splanchnic nerve.

Spleen [splēn] (Gr. *splen* = spleen).
A large lymphoid organ located near the left side of the stomach. In various vertebrates and at different times in their life cycle, it produces, stores, or disassembles senescent red blood cells.

Splenius muscle [splē'nē-ŭs] (Gr. *splenion* = bandage).
A thin, triangular sheet of muscle located on the back of the neck of mammals deep to part of the rhomboideus muscle.

Stapes [stā'pēz] (L. = stirrup).
The single auditory ossicle of nonmammalian vertebrates, and the innermost of the three auditory ossicles in mammals.

Statoacoustic nerve [stat'ō-ă-kū'stik] (Gr. *statos* = standing still + *akoustikos* = pertaining to hearing).
A name often used for the eighth cranial nerve in nonmammalian vertebrates. It is called the vestibulocochlear nerve in mammals.

Sternum [ster'nŭm] (Gr. *sternon* = chest).
The breastbone of terrestrial vertebrates.

Stomach [stŭm'ŭk] (Gr. *stomakos* = stomach).
Saclike part of the digestive tract lying between the esophagus and intestine in which food is temporarily stored and digestion usually is initiated.

Stratum [strat'ŭm] (L. = layer).
A layer of tissue, such as the stratum corneum of the skin.

Styloid process [stī'loyd] (Gr. *stylos* = pillar + *eidos* = resembling).
A slender process on the ventral surface of the skull of many mammals. The hyoid apparatus attaches to it.

Subclavian [sŭb-klā'vē-an] (L. *sub-* = beneath + *clavis* = key).
Pertains to a position beneath the clavicle, e.g., the subclavian artery.

Sublingual gland [sŭb-ling'gwăl] (L. *lingua* = tongue).
A salivary gland of mammals. It lies beneath the tongue.

Submucosa [sŭb-mŭ-kō'să] (L. *mucosus* = mucous).
A layer of vascularized connective tissue in the wall of the digestive or respiratory tract. It underlies the mucosa.

Sulcus [sŭl'kŭs] (L. = groove).
A groove on the surface of an organ, e.g., the sulci on the surface of the cerebrum of many mammals.

Superior [sū-pēr'ē-ōr] (L. *super* = above + *-ior* = suffix indicating the comparative form of an adjective, e.g., more above).
A direction toward the head of a human being.

Suprarenal gland [sū-pră-rē'năl].
See **Adrenal gland.**

Suture [sū'chūr] (L. = seam).
An essentially immovable joint in which the bones are separated by connective tissue, e.g., many of the joints between bones of the skull.

Sympathetic nervous system [sim-pă-thet'ik] (Gr. *syn* = with + *pathos* = feeling).
The part of the autonomic nervous system that, in mammals, leaves the central nervous system from the thoracic and lumbar parts of the spinal cord. Its activity helps an animal adjust to stress by promoting physiological processes that mobilize the energy available to the tissues.

Symphysis [sim'fi-sis] (Gr. *physis* = a growth).
A joint or fusion between bones that permits limited movement by the deformation of fibrocartilage between them. It occurs in the midline of the body, e.g., the pelvic symphysis, mandibular symphysis.

Systemic [sis-tem'ik] (Gr. *systema* = whole).
Pertaining to the body as a whole rather than to a specific part, e.g., the systemic circulation as opposed to the pulmonary circulation.

Talus [tā'lus] (L. = ankle bone).
The proximal tarsal bone of mammals. It articulates with the tibia.

Tapetum lucidum [tă-pē'tŭm lū'sid-ŭm] (L. = carpet + *lucidus* = shining).
A layer within or behind the retina of some vertebrates. It reflects light back onto the photoreceptive cells.

Tarsal [tar'săl] (Gr. *tarsos* = ankle).
One of the small bones of the ankle.

Tela choroidea [tē'la kōr-oy'dē-a] (L. *tela* = web, Gr. *chorion* = membrane encasing the fetus + *eidos* = resembling).
A thin membrane forming the roof or part of the wall of some ventricles of the brain. It is composed of the ependymal epithelium and the pia mater.

Telencephalon [tel-en-sef'ă-lon] (Gr. *telos* = end + *enkephalos* = brain).
The most rostral of the five brain regions. It includes the olfactory lobes and cerebrum.

Temporal [tem'pŏ-răl] (L. *tempus*, gen. *temporis* = time).
Pertaining to the temporal region of the skull, so called because the hair in this region is the first to become gray in human beings.

Temporal fossa.
A depression on the lateral surface of the mammalian skull caudal to the orbit and dorsal to the zygomatic arch. It lodges the temporal jaw muscle.

Tendon [ten'dŏn] (L. *tendo* = to stretch).
A band of dense connective tissue attaching a muscle to a bone, or sometimes to another muscle.

Tendon of Achilles (*Achilles*, the hero of Homer's *Iliad* was said to be invulnerable except for this tendon).
The tendon that extends from the large muscle mass on the caudal surface of the crus to the calcaneus. These muscles are powerful extensors of the foot.

Tentorium [ten-tō'rē-um] (L. = tent, from *tendo*, p. p. *tentum* = to stretch).
A septum of dura mater in mammals located between the cerebrum and cerebellum. It ossifies in some species.

Teres [ter'ēz] (L. = rounded, smooth).
Descriptive of a round structure, such as the teres muscles.

Tertiary follicle.
A mature follicle in the ovary; also called a *Graafian follicle.*

Testis [tes'tis] (L. = witness, in ancient Rome necessarily an adult male).
The male gonad. It produces sperm cells and the hormone testosterone.

Tetrapod [tet'ră-pod] (Gr. *tetra* = four + *pous*, gen. *podos* = foot).
A collective term for terrestrial vertebrates. They have four feet unless some have been secondarily lost or modified.

Thalamus [thal'ă-mus] (Gr. *thalamos* = inner chamber, bedroom).
The lateral walls of the diencephalon. It is an important center between the cerebrum and other parts of the brain.

Theca [thē'kă] (Gr. *theke* = box, case).
A case or covering, such as the group of connective tissue cells that form the outer layer of an ovarian follicle.

Thoracolumbar fascia [thōr'ă-kō-lŭm'bar].
A part of the deep fascia on the dorsal surface of the thoracic and lumbar regions. It is associated with the muscles of the back and abdominal wall.

Thorax [thō'raks] (Gr. = chest).
The region of the mammalian body encased by the ribs and sternum.

Thymus [thī'mŭs] (Gr. *thymos* = thyme, thymus; so called because of its resemblance to a bunch of thyme).
A lymphoid organ in the ventral part of the neck and thorax. It is essential for the maturation of T-lymphocytes and probably other parts of the immune system. The gland is best developed in young individuals and atrophies later in life.

Thyroid gland [thī'royd] (Gr. *thyreos* = oblong shield + *eidos* = resembling).
An endocrine gland that is usually located near the cranial end of the trachea, but lies over the thyroid cartilage of the larynx in human beings (whence its name). Its hormones increase the rate of metabolism.

Tibia [tib'ē-ă] (L. = the large shinbone).
The large bone on the medial side of the lower leg.

Tissue [tish'ū] (Old French *tissu* = cloth).
An aggregation of cells that together perform a common function.

Tongue [tŭng] (Old English *tunge* = tongue).
A muscular organ in the floor of the oral cavity that often helps gather food and manipulates it within the mouth cavity.

Tonsil [ton'sil] (L. *tonsilla* = tonsil).
One of the lymphoid organs that develop in the wall of the mammalian pharynx.

Trachea [trā'kē-ă] (Gr. *tracheia* = rough artery).
The respiratory tube between the larynx and the bronchi.

Tract [trakt] (L. *traho*, p.p. *tractus* = to pull).
(1) A linear group of organs having a similar function, e.g., the digestive tract. (2) A group of axons of similar function traveling together in the central nervous system.

Transverse [trans-vers'] (L. *transversus* = transverse).
A plane of the body crossing its longitudinal axis at right angles.

Transverse process.
A process of a vertebra, that lies in the transverse plane. Either the tubercle of a rib articulates with it, or an embryonic rib becomes incorporated in it and serves as an attachment site for muscles.

Trapezoid body [trap'ĕ-zōyd] (Gr. *trapezoides* = resembling a trapezium).
An acoustic commissure at the rostral end of the ventral surface of the mammalian medulla oblongata.

Trigeminal nerve [trī-jem'i-năl] (L. *trigeminus* = threefold).
The fifth cranial nerve. It has three branches in mammals. It innervates the jaw muscles and returns sensory fibers from the surface of the head and oral cavity, except for taste buds.

Trochanter [trō-kan'ter] (Gr. = a runner).
One of the processes on the proximal end of the femur to which certain pelvic and thigh muscles attach.

Trochlear nerve [trok'lē-ăr] (L. *trochlea* = pulley).
The fourth cranial nerve. It innervates one of the extrinsic muscles of the eye, which, in mammals, passes through a connective tissue pulley before inserting on the eyeball.

Truncus arteriosus [trŭng'kŭs ar-ter'ē-ō-sus] (L. = trunk, stem).
One of two arterial trunks in the frog leading from the cranial end of the heart to arterial arches supplying the skin and lungs, head, and body.

Tunic [tū'nik] (L. *tunica* = a coat, covering).
Descriptive of a layer of an organ, such as one of the layers of the eyeball.

Tunica albuginea [tū'ni-kă al-byū-jin'ē-ă] (L. *albugineus*, from *albugo* = white spot).
A white, fibrous capsule, such as the one forming the wall of the testis and sending septa into the testis.

Tympanic [tim-pan'ik] (L. *tympanum* = drum).
Pertaining to the middle ear.

Tympanic cavity.
The cavity of the middle ear that lies between the tympanic membrane and the inner ear within the otic capsule. One or three auditory ossicles traverse it, and the auditory tube connects it with the pharynx.

Tympanic membrane.
The eardrum.

Ulna [ŭl'nă] (L. = elbow bone).
One of the bones of the antebrachium of terrestrial vertebrates. It extends behind the elbow in mammals and lies on the medial side of the antebrachium when the hand is supine.

Umbilical [ŭm-bil'i-kăl] (L. *umbilicus* = navel).
Pertaining to the navel, e.g., umbilical cord, umbilical artery.

Ureter [yū-rē'ter] (Gr. *oureter* = ureter, from *ouron* = urine).
The duct in amniotes that carries urine from the kidney to the urinary bladder.

Urethra [yū-rē'thră] (Gr. *ourethra* = urethra).
The duct that carries urine from the urinary bladder to the cloaca or to the outside of the body in amniotes; part of it also carries spinal fluid and sperm cells in males.

Urinary bladder [yūr'i-nār-ē] (Gr. *ouron* = urine).
A saccular organ in terrestrial vertebrates in which urine from the kidney accumulates before being discharged from the body.

Urogenital [yū'rō'jen-i-tăl] (L. *genitalis* = creative, fruitful).
A combining term for the urinary and genital systems, which share many ducts.

Urostyle [yū'rō'stīl] (Gr. *oura* = tail + *stylos* = a pillar, peg).
The rodlike, caudal part of the vertebral column of a frog. It is composed of several fused caudal vertebrae.

Uterine tube [yū'ter-in] (L. *uterus* = womb).
One of a pair of narrow tubes in mammals. It extends from the vicinity of the ovary to the uterus and carries fertilized eggs to the uterus. Also called the *Fallopian tube.*

Uterus [yū'ter-ŭs].
The organ in females in which embryos develop in live-bearing species. It develops from part of the oviduct.

Utriculus [yū'trik'yū-lŭs] (L. = small sac).
The upper chamber of the inner ear to which the semicircular ducts attach.

Vagina [vă-jī'na] (L. = sheath).
The part of the mammalian female reproductive tract that receives the penis during copulation.

Vaginal cavity.
The part of the coelom that contains the testis of male mammals.

Vaginal vestibule [ves'ti-būl] (L. *vestibulum* = antechamber).
The passage or space into which the vagina and urethra enter in female mammals. It is very shallow in human beings but often forms a short canal in quadrupeds.

Vagus nerve [vā'gŭs] (L. = wandering).
The tenth cranial nerve. It carries motor fibers to muscles of the larynx and parasympathetic fibers to thoracic and abdominal organs; it returns sensory fibers from these parts of the body.

Vasa efferentia.
See **Efferent ductules.**

Vascular tunic.
The middle layer of the eyeball. It forms the choroid, ciliary body, and iris.

Vas deferens.
See **Ductus deferens.**

Vastus [vas'tus] (L. = great, large).
Descriptive of some thigh muscles, e.g., the vastus lateralis.

Vein [vān] (L. *vena* = vein).
A blood vessel that carries blood toward the heart. Usually the blood is low in oxygen content, but pulmonary veins from the lungs have blood with a high oxygen content.

Vena cava [vē'nă cā'vă] (L. = hollow vein).
One of the primary veins of frogs and amniotes. It leads directly to the heart.

Ventral [ven'tral] (L. *ventralis* = ventral, from *venter* = belly).
A direction toward the underside of a quadruped.

Ventricle [ven'tri-kl] (L. *ventriculus* = small belly).
(1) A chamber of the heart. It greatly increases blood pressure and sends the blood to the lungs or to the body. (2) One of the cavities within the brain.

Vermis [ver'mis] (L. = worm).
The "segmented" and wormlike median portion of the mammalian cerebellum.

Vertebra [ver'te-bră] (L. = vertebra, joint).
One of the units that make up the vertebral column.

Vertebral arch.
The arch of a vertebra that surrounds the spinal cord. It is also called a *neural arch.*

Vertebral body.
The main, supporting component of a vertebra. It lies ventral to the vertebral arch. Also called a vertebral centrum.

Vesicular gland [vĕ-sik'yū-lăr] (L. *vesicula* = small bladder).
One of the accessory genital glands of male mammals that contributes to the seminal fluid; also called the *seminal vesicle.*

Vestibulocochlear nerve [ves-tib'yū-lō-kok-lē-ăr] (L. *cochlea* = snail shell).
The eighth cranial nerve. It returns sensory fibers from the parts of the inner ear related to equilibrium (vestibular apparatus) and from the part monitoring sound detection (cochlea). Often called the statoacoustic nerve in anamniotes, in which a cochlea is absent.

Vibrissae [vī-bris'ē] (L. = vibrissae, from *vibro* = to quiver).
Long, tactile hairs on the snout of many mammals.

Villi [vil'i] (L. = shaggy hair).
Multicellular, but minute, often finger-shaped projections of an organ. They increase its surface area, e.g., the intestinal villi.

Visceral [vis'er-ăl] (L. *viscus*, pl. *viscera* = internal organs).
Pertaining to the inner part of the body as opposed to the body wall and appendages, e.g., visceral muscles, visceral skeleton.

Visceral arches.
The skeletal arches that develop in the wall of the pharynx and may contribute to the formation of the jaws (frog) and parts of the skull, hyoid, and larynx.

Vitreous body [vit'rē-ŭs] (L. *vitreus* = glassy).
The clear, viscous material in the eyeball between the lens and retina.

Viviparous [vī-vip'ă-rŭs] (L. *vivus* = living + *pario* = bearing).
Live-bearing. A pattern of reproduction found in most mammals and a few other vertebrates in which the young develop in a uterus and are born fully formed.

Vocal cords [vō'kăl] (L. *vocalis* = pertaining to voice).
Folds of mucous membrane within the larynx of frogs and mammals. Involved in the production of sounds.

Vomer bone [vō'mer] (L. = plowshare).
A bone in the roof of the mouth in frogs and in the floor of the nasal cavities in mammals. It is plowshare-shaped.

Vomeronasal organ [vō'mer-ō-nā-zăl].
An accessory olfactory organ located between the palate and the nasal cavities of most terrestrial vertebrates. It is important in feeding and sexual behaviors. It is also called **Jacobson's organ.**

Vulva [vŭl'vă] (L. = covering).
The female external genitalia.

Wharton's jelly (*Thomas Wharton*, British anatomist and physician, 1614 - 1673).
The mucoid connective tissue of the umbilical cord.

White matter.
Tissue in the central nervous system that consists primarily of myelinated axons.

Wolffian duct (*Kaspar Friedrich Wolff*, German embryologist in Russia, 1733–1794).
See **Archinephric duct.**

Yolk sac [yōk] (Old English *geulca* = yolk, from *geolu* = yellow).
The yolk-containing sac attached to the ventral surface of the embryo in some fishes, reptiles, birds, and egg-laying mammals. It is reduced in viviparous mammals.

Zonule fibers [zō'nyūl] (L. *zonula* = small zone).
Delicate fibers extending between the ciliary body and the lens equator. These fibers transmit forces from the ciliary body to the lens.

Zygomatic arch [zī'gō-mat-ik] (Gr. *zygoma* = bar, yoke).
The arch of bone beneath the orbit in a mammalian skull. It connects the facial and cranial regions.